2018 SQA Past Papers with Answers

Advanced Higher
PHYSICS

2016, 2017 & 2018 Exams

Hodder Gibson Study Skills Advice – Advanced Higher Physics	– page 3
Hodder Gibson Study Skills Advice – General	– page 7
2016 EXAM	– page 9
2017 EXAM	– page 61
2018 EXAM	– page 113
ANSWERS	– page 165

AN HACHETTE UK COMPANY

This book contains the official 2016, 2017 and 2018 Exams for Advanced Higher Physics, with associated SQA-approved answers modified from the official marking instructions that accompany the paper.

In addition the book contains study skills advice. This advice has been specially commissioned by Hodder Gibson, and has been written by experienced senior teachers and examiners in line with the Advanced Higher syllabus and assessment outlines. This is not SQA material but has been devised to provide further guidance for Advanced Higher examinations.

Hodder Gibson is grateful to the copyright holders for permission to use their material. Every effort has been made to trace the copyright holders and to obtain their permission for the use of copyright material. Hodder Gibson will be happy to receive information allowing us to rectify any error or omission in future editions.

Permission has been sought from all relevant copyright holders and Hodder Gibson is grateful for the use of the following:

Image © Calvin Chan/Shutterstock.com (2016 page 3);
Image © Diego Barbieri/Shutterstock.com (2017 page 3);
Image © Vasca/stock.adobe.com (2017 page 34);
Image © CG Stocker/Shutterstock.com (2018 page 3);
Image © dashadima/Shutterstock.com (2018 page 8);
Image © Luciano Cosmo/Shutterstock.com (2018 page 38).

Hachette UK's policy is to use papers that are natural, renewable and recyclable products and made from wood grown in sustainable forests. The logging and manufacturing processes are expected to conform to the environmental regulations of the country of origin.

Orders: please contact Bookpoint Ltd, 130 Park Drive, Milton Park, Abingdon, Oxon OX14 4SE. Telephone: (44) 01235 827827. Fax: (44) 01235 400454. Lines are open 9.00–5.00, Monday to Saturday, with a 24-hour message answering service. Visit our website at www.hoddereducation.co.uk. Hodder Gibson can also be contacted directly at hoddergibson@hodder.co.uk

This collection first published in 2018 by
Hodder Gibson, an imprint of Hodder Education,
An Hachette UK Company
211 St Vincent Street
Glasgow G2 5QY

Advanced Higher 2016, 2017 and 2018 Exam Papers and Answers © Scottish Qualifications Authority. Study Skills section © Hodder Gibson. All rights reserved. Apart from any use permitted under UK copyright law, no part of this publication may be reproduced or transmitted in any form or by any means, electronic or mechanical, including photocopying and recording, or held within any information storage and retrieval system, without permission in writing from the publisher or under licence from the Copyright Licensing Agency Limited. Further details of such licences (for reprographic reproduction) may be obtained from the Copyright Licensing Agency Limited, www.cla.co.uk.

Typeset by Aptara, Inc.

Printed in the UK

A catalogue record for this title is available from the British Library

ISBN: 978-1-5104-5552-8

2 1

2019 2018

Introduction
Advanced Higher Physics

The course
The course comprises two whole and two half units:
- Rotational Motion and Astrophysics (1 unit)
- Quanta and Waves (1 unit)
- Electromagnetism (0·5 unit)
- Investigating Physics (0·5 unit)

Course assessment
In order to gain an award in the course, you must pass the unit assessment in each topic. The first three topics will be internally assessed by your teacher/lecturer and the evidence externally verified. The assessments will be in the form of class tests.

Investigating Physics
The Investigating Physics unit assessment will involve you planning your experiments and taking note of your experimental readings in your record of work/diary.

Ensure this record is neat and tidy so that when you come to type your report, it is easy to decipher. Your teacher will overview and sign your record of work/diary.

This record could be externally verified by SQA and should be submitted to your teacher along with your project write up.

The examination
This will last 2·5 hours and will be marked out of **140 marks** and then scaled back to **100 marks**.

The project
The report will be submitted to SQA for marking and will be marked out of **30 marks**.

This will then be added to the examination mark giving a total out of **130 marks**.

Experimental write ups should be typed up as you progress through the course. This will build up your skills and also allow you to easily go back and correct any mistakes or take on any suggestions from your teacher.

Become confident in the use of software packages that allow you to plot graphs including error bars. Find out how to find the gradient of a straight line and its associated uncertainty – the use of Linest function in the Microsoft package. This will save you an enormous amount of time!

See the SQA website for advice to candidates in relation to the project.

Advanced Higher Physics is an excellent qualification and although it is a step above Higher, the fact that you are in the class proves you have the ability to do well. It just takes hard work and application – the danger is to have too many distractions in your sixth year. Try to choose a project that suits and **stick to deadlines**!

Examination tips
Mark allocation
- **4 or 5 marks** will generally involve more than one step or several points of coverage.
- **3 marks** will generally involve the use of just one equation.

Standard 3-mark example
Calculate the acceleration of a mass of 5 kg when acted on by a resultant force of 10 N.

Solution 1	Solution 2	Solution 3
$F = ma$ **(1)**	$F = ma$ **(1)**	$F = ma$ **(1)**
$10 = 5a$ **(1)**	$a = \dfrac{m}{F} = \dfrac{5}{10}$ **(0)**	$10 = 5a$ **(1)**
$a = 2\,\text{ms}^{-2}$ **(1)**	$= 0{\cdot}5\,\text{ms}^{-2}$	$a = 0{\cdot}5\,\text{ms}^{-2}$
3 marks	**1 mark for selecting formula.**	**2 marks for selecting formula + correct substitution.**

Do not rearrange equations in algebraic form. Select the appropriate equation, substitute the given values then rearrange the equation to obtain the required unknown. This minimises the risk of a wrong substitution.

Use of the Data Sheet
Clearly show where you have substituted a value from the data sheet. For example, **do not** leave μ_0 in an equation. You must show the substitution $4\pi \times 10^{-7}$ in your equation.

When rounding, **do not** round the given values. For example, the mass of a proton = $1{\cdot}673 \times 10^{-27}$ kg **not** $1{\cdot}67 \times 10^{-27}$ kg.

Use of the Physics Relationship Sheet
Although many of the required equations are given, it is better to know the basic equations to gain time in the examination.

"Show" questions
Generally **all steps** for these must be given, even though they might seem obvious. **Do not assume that substitutions are obvious to the marker.**

All equations used must be stated separately and then clearly substituted if required. Many candidates will look at the end product and somehow end up with the required answer. The marker has to ensure that the path to the solution is clear. It is good practice to state why certain equations are used, explaining the Physics behind them.

Definitions
Know and understand definitions given in the course. Definitions often come from the interpretation of an equation.

Diagrams

Use a ruler and the appropriate labels. Angles will be important in certain diagrams. Too many candidates attempt to draw ray diagrams freehand.

Graphs

Read the question and ensure you know what is being asked. Label your graph correctly and do not forget to label the origin.

"Explain" and "describe" questions

These tend to be done poorly.

- Ensure all points are covered and read over again in order to check there are no mistakes. Try to be clear and to the point, highlighting the relevant Physics.
- Do not use up and down arrows in a description – this may help you in shorthand, but these must be translated to words.
- Be aware some answers require justification. No attempt at a justification can mean no marks awarded.

Two or more attempts at an answer

Any attempt that you do not want the marker to consider should be scored out. Otherwise zero marks could be awarded.

Do not be tempted to give extra information that might be incorrect – all the marks may not be awarded if there are incorrect pieces of information. For example, this might include converting **incorrectly** m to nm in the last line of an answer, when it is not required.

At the end of the exam, if you have time, quickly go over each answer and make sure you have the correct unit inserted.

Experimental descriptions/planning

You could well be called on to describe an experimental set-up. Ensure your description is clear enough for another person to repeat it, and include a clearly labelled diagram.

Suggested improvements to the experimental procedure

Look at the percentage uncertainties in the measured quantities and decide which is most significant. Suggest how the size of this uncertainty could be reduced – do not suggest using better apparatus! It might be better to repeat readings, so that random uncertainty is reduced or increase distances to reduce the percentage uncertainty in scale reading. There could be a systematic uncertainty that is affecting all readings. It really depends on the experiment. Use your judgement.

Handling data

Relationships

There are two methods to prove a direct or inverse relationship.

Graphical approach

Plot the graphs with the appropriate x and y values and look for a straight line – better plotted in pencil in case of mistakes. **Do not force a line through the origin!**

(A vs B for a direct relationship, C vs $\frac{1}{D}$ for an inverse relationship)

> Using the equation of a straight line,
> **y = mx + c.**
> Be aware that the gradient of the line can often lead to required values.

Example: Finding the permeability of free space.

$B = \frac{\mu_0 I}{2\pi r}$

Express $B = \frac{\mu_0}{2\pi r} I$ in the form of y = mx + c (c = 0)

By plotting the graph of B against I, the value of the gradient will give $\frac{\mu_0}{2\pi r}$

(Note: the line should not be forced through the origin.)

The value of C can often indicate a systematic uncertainty in the experiment. Ensure you are clear on how to calculate the gradient of a line.

Algebraic approach

If it appears that A ∝ B then calculate the value of $\frac{A}{B}$ **for all values**.

If these show that $\frac{A}{B}$ = k then the relationship holds.

If it appears that C ∝ $\frac{1}{D}$ then calculate the value C.D **for all values**.

If these show that C.D = k then the relationship holds.

Uncertainties

In this area, you must understand the following:

- systematic, calibration, scale reading (analogue and digital) and random uncertainties
- percentage/fractional uncertainties
- combinations – Pythagorean relationship
- absolute uncertainty in the final answer (give to one significant figure).

For your lab work, it is always useful and less time consuming to use a spreadsheet package to shortcut calculations, plot graphs and estimate uncertainties. Just ensure that if plotting a graph, it is at least half a page in size and use the smallest grid lines available.

Significant figures

Do not round off in intermediate calculations, but round off in the final answer to an appropriate number of figures. **Rounding off to three significant figures in the final answer will generally be acceptable.**

Prefixes

Ensure you know all the required prefixes and be able to convert them to the correct power of 10.

Open-ended questions

There will be two open-ended questions in the paper each worth 3 marks. Some candidates look upon these as mini essays.

Remember each is worth only 3 marks and they give the opportunity to demonstrate knowledge and understanding. **However, do not spend too long on these.** It might be better to revisit them at the end of the exam. Some students prefer to use bullet points to highlight the main areas of understanding.

Below is some advice on each unit. Obviously, all points in all units cannot be covered, but hopefully the following can give you a start in what to look out for.

Unit 1 Rotational Motion and Astrophysics

Kinematic relationships

Calculus — equations of motion

$$s = f(t) - \text{given}$$

Be clear that differentiating once gives the velocity, differentiating twice gives the acceleration.

$$a = f(t) - \text{given}$$

Integrating once gives the velocity, integrating twice gives the displacement. Remember to take into account the constant of integration each time by considering the limits or initial conditions.

Circular Motion

Make a two column table with the headings "Linear" and "Circular". On the left-hand side, write down all the equations you have come across in Higher, starting with speed and then the equations of motion. On the right-hand side, add all the equivalent equations for circular motion.

Linear	Circular
$v = \frac{s}{t}$	$\omega = \frac{\theta}{t}$
$v = u + at$	$\omega = \omega_0 + \alpha t$

Add to these as you progress through the course.

Central force

There will often be a question on **central** or **centripetal force**. Remember this is the force that acts on an object causing the object to follow a circular path.

There is no outward (centrifugal) force acting on that object and it would be good advice **not to mention this force** in your description.

It is worth noting that a fun fair ride might give the impression of feeling an outward force acting on a person, but this is an unreal force – they just think the force is outward. In fact, it is the inward force from the seat that enables you to follow the circular path.

The same will apply with a "loop the loop" ride where, at the top of the loop, the track and weight of the car supplies the required centripetal force.

Moment of inertia can be defined as a measure of the resistance to angular acceleration about a given axis (resistance to change).

Gravitation

Circular motion and planetary motion

The key to calculating the period, T, of motion of a planet or satellite lies with gravitational force supplying the central force, that is:

$$\frac{GMm}{r^2} = mr\omega^2 \text{ then use } \omega = \frac{2\pi}{T} \text{ to find } T.$$

Escape velocity derivation

The starting point is the realisation when the object has escaped the pull of gravity **then $E_T = 0$**.

For example, $E_T = E_k + E_p = \frac{1}{2}mv^2 - \frac{GMm}{r} = 0.$

Rearrange the equation to obtain the **expression for v**.

$\left(\text{Note that } E_p = -\frac{GMm}{r}\right)$

General relativity

Remember the higher the altitude, then the gravitational field will be less, **which means clocks will run fast**. GPS systems have to take this into account, otherwise the position determined on Earth would be incorrect.

Mass curves spacetime which causes gravitational pull – rubber sheet analogy.

Stellar Physics

Become familiar with the Hertzsprung-Russell (H-R) diagram – stellar evolution.

Unit 2 Quanta and Waves

Bohr model of the atom

Many candidates omit the fact that the angular momentum of the electrons is quantised.

Wave particle duality

Experimental evidence is often confused.

- Electron diffraction through crystals – particles as waves.
- Photoelectric effect – waves as particles.

Uncertainty principle

It is difficult to measure something precisely without the measuring procedure affecting what you are trying to measure!

Simple harmonic motion: know the definition of SHM

Be able to derive the velocity (differentiate once) and acceleration (differentiate again) given the expression for displacement:

$$y = A\sin\omega t \text{ or } x = A\cos\omega t$$

From v and a the expressions for the maximum velocity and maximum acceleration can be found. (These occur when the maximum value of cos or sin = 1.)

Waves

Wave Equation

$$y = A\sin 2\pi\left(ft - \frac{x}{\lambda}\right)$$

Just knowing the coefficients of t and x in the wave equation allows the retrieval of frequency, wavelength and the speed of the wave,

$$\text{Coefficient of } t = 2\pi f$$
$$\text{Coefficient of } x = \frac{2\pi}{\lambda}$$

Remember the 2π!

Standing or Stationary waves

Remember these are produced when a reflected wave interferes with the incident wave causing maxima (antinodes) and minima (nodes).

Unit 3 Electromagnetism

Electric field strength

Parallel plates   **Point charge**

(uniform field between plates)

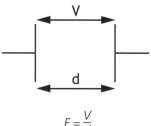
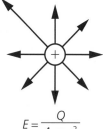

$$E = \frac{V}{d} \qquad\qquad E = \frac{Q}{4\pi\varepsilon_0 r^2}$$

Candidates often mix up these formulae for electric field strength. They are quite different!

Ferromagnetism

Certain materials can become magnetised due to the alignment of their magnetic dipoles.

Charged particles in a magnetic field

A circular orbit will be produced if the charged particle cuts the field at 90°.

The central force is produced by the magnetic force acting on the particle.

$$\frac{mv^2}{r} = Bqv$$

If the charge enters the field at an angle **less than 90°** then the resultant path will be **helical**.

An example of this would be charged particles being deflected by the Earth's magnetic field to the North or South poles producing the borealis.

Candidates should be able to explain why the resulting motion is helical.

Treat this as two components of velocity:

V_H = constant and moves in the direction of field lines – no force acting.

V_V = 90° to the field which produces a central force causing the charge particle to follow a circular path.

Back emf in an inductor (coil)

$$E = -L\frac{dI}{dt}$$

Remember that the back emf is produced by a changing current, which in turn produces a **changing magnetic field** throughout the coil. Many candidates omit this in their explanation.

Be aware that if asked to calculate the inductance L (or the rate of change of current) and a **negative answer** is obtained then **there has been an error. Invariably the fact that the induced emf is negative has not been taken into account**.

For example:

$$E = 9V \frac{dI}{dt} = 5\,As^{-1}$$
$$E = -L\frac{dI}{dt}$$
$$9 = -L \times 5$$
$$L = -1.8\,H$$

This is incorrect!

$$E = -9V \frac{dI}{dt} = 5\,As^{-1}$$
$$E = -L\frac{dI}{dt}$$
$$-9 = -L \times 5$$
$$L = 1.8\,H$$

This is correct!

Study Skills – what you need to know to pass exams!

General exam revision: 20 top tips

When preparing for exams, it is easy to feel unsure of where to start or how to revise. This guide to general exam revision provides a good starting place, and, as these are very general tips, they can be applied to all your exams.

1. Start revising in good time.

Don't leave revision until the last minute – this will make you panic and it will be difficult to learn. Make a revision timetable that counts down the weeks to go.

2. Work to a study plan.

Set up sessions of work spread through the weeks ahead. Make sure each session has a focus and a clear purpose. What will you study, when and why? Be realistic about what you can achieve in each session, and don't be afraid to adjust your plans as needed.

3. Make sure you know exactly when your exams are.

Get your exam dates from the SQA website and use the timetable builder tool to create your own exam schedule. You will also get a personalised timetable from your school, but this might not be until close to the exam period.

4. Make sure that you know the topics that make up each course.

Studying is easier if material is in manageable chunks – why not use the SQA topic headings or create your own from your class notes? Ask your teacher for help on this if you are not sure.

5. Break the chunks up into even smaller bits.

The small chunks should be easier to cope with. Remember that they fit together to make larger ideas. Even the process of chunking down will help!

6. Ask yourself these key questions for each course:

- Are all topics compulsory or are there choices?
- Which topics seem to come up time and time again?
- Which topics are your strongest and which are your weakest?

Use your answers to these questions to work out how much time you will need to spend revising each topic.

7. Make sure you know what to expect in the exam.

The subject-specific introduction to this book will help with this. Make sure you can answer these questions:

- How is the paper structured?
- How much time is there for each part of the exam?
- What types of question are involved? These will vary depending on the subject so read the subject-specific section carefully.

8. Past papers are a vital *revision tool!*

Use past papers to support your revision wherever possible. This book contains the answers and mark schemes too – refer to these carefully when checking your work. Using the mark scheme is useful; even if you don't manage to get all the marks available first time when you first practise, it helps you identify how to extend and develop your answers to get more marks next time – and of course, in the real exam.

9. Use study methods that work well for you.

People study and learn in different ways. Reading and looking at diagrams suits some students. Others prefer to listen and hear material – what about reading out loud or getting a friend or family member to do this for you? You could also record and play back material.

10. There are three tried and tested ways to make material stick in your long-term memory:

- Practising – e.g. rehearsal, repeating
- Organising – e.g. making drawings, lists, diagrams, tables, memory aids
- Elaborating – e.g. incorporating the material into a story or an imagined journey

11. Learn actively.

Most people prefer to learn actively – for example, making notes, highlighting, redrawing and redrafting, making up memory aids, or writing past paper answers. A good way to stay engaged and inspired is to mix and match these methods – find the combination that best suits you. This is likely to vary depending on the topic or subject.

12. Be an expert.
Be sure to have a few areas in which you feel you are an expert. This often works because at least some of them will come up, which can boost confidence.

13. Try some visual methods.
Use symbols, diagrams, charts, flashcards, post-it notes etc. Don't forget – the brain takes in chunked images more easily than loads of text.

14. Remember – practice makes perfect.
Work on difficult areas again and again. Look and read – then test yourself. You cannot do this too much.

15. Try past papers against the clock.
Practise writing answers in a set time. This is a good habit from the start but is especially important when you get closer to exam time.

16. Collaborate with friends.
Test each other and talk about the material – this can really help. Two brains are better than one! It is amazing how talking about a problem can help you solve it.

17. Know your weaknesses.
Ask your teacher for help to identify what you don't know. Try to do this as early as possible. If you are having trouble, it is probably with a difficult topic, so your teacher will already be aware of this – most students will find it tough.

18. Have your materials organised and ready.
Know what is needed for each exam:
- Do you need a calculator or a ruler?
- Should you have pencils as well as pens?
- Will you need water or paper tissues?

19. Make full use of school resources.
Find out what support is on offer:
- Are there study classes available?
- When is the library open?
- When is the best time to ask for extra help?
- Can you borrow textbooks, study guides, past papers, etc.?
- Is school open for Easter revision?

20. Keep fit and healthy!
Try to stick to a routine as much as possible, including with sleep. If you are tired, sluggish or dehydrated, it is difficult to see how concentration is even possible. Combine study with relaxation, drink plenty of water, eat sensibly, and get fresh air and exercise – all these things will help more than you could imagine. Good luck!

ADVANCED HIGHER
2016

National Qualifications 2016

X757/77/11

**Physics
Relationships Sheet**

TUESDAY, 24 MAY
9:00 AM – 11:30 AM

Relationships required for Physics Advanced Higher

$v = \dfrac{ds}{dt}$

$a = \dfrac{dv}{dt} = \dfrac{d^2s}{dt^2}$

$v = u + at$

$s = ut + \dfrac{1}{2}at^2$

$v^2 = u^2 + 2as$

$\omega = \dfrac{d\theta}{dt}$

$\alpha = \dfrac{d\omega}{dt} = \dfrac{d^2\theta}{dt^2}$

$\omega = \omega_o + \alpha t$

$\theta = \omega_o t + \dfrac{1}{2}\alpha t^2$

$\omega^2 = \omega_o^2 + 2\alpha\theta$

$s = r\theta$

$v = r\omega$

$a_t = r\alpha$

$a_r = \dfrac{v^2}{r} = r\omega^2$

$F = \dfrac{mv^2}{r} = mr\omega^2$

$T = Fr$

$T = I\alpha$

$L = mvr = mr^2\omega$

$L = I\omega$

$E_K = \dfrac{1}{2}I\omega^2$

$F = G\dfrac{Mm}{r^2}$

$V = -\dfrac{GM}{r}$

$v = \sqrt{\dfrac{2GM}{r}}$

apparent brightness, $b = \dfrac{L}{4\pi r^2}$

Power per unit area $= \sigma T^4$

$L = 4\pi r^2 \sigma T^4$

$r_{Schwarzschild} = \dfrac{2GM}{c^2}$

$E = hf$

$\lambda = \dfrac{h}{p}$

$mvr = \dfrac{nh}{2\pi}$

$\Delta x\, \Delta p_x \geq \dfrac{h}{4\pi}$

$\Delta E\, \Delta t \geq \dfrac{h}{4\pi}$

$F = qvB$

$\omega = 2\pi f$

$a = \dfrac{d^2y}{dt^2} = -\omega^2 y$

$y = A\cos\omega t$ or $y = A\sin\omega t$

$v = \pm\omega\sqrt{(A^2 - y^2)}$

$E_K = \frac{1}{2}m\omega^2(A^2 - y^2)$

$E_P = \frac{1}{2}m\omega^2 y^2$

$y = A\sin 2\pi(ft - \frac{x}{\lambda})$

$E = kA^2$

$\phi = \frac{2\pi x}{\lambda}$

optical path difference $= m\lambda$ or $\left(m + \frac{1}{2}\right)\lambda$

where $m = 0, 1, 2....$

$\Delta x = \frac{\lambda l}{2d}$

$d = \frac{\lambda}{4n}$

$\Delta x = \frac{\lambda D}{d}$

$n = \tan i_P$

$F = \frac{Q_1 Q_2}{4\pi\varepsilon_o r^2}$

$E = \frac{Q}{4\pi\varepsilon_o r^2}$

$V = \frac{Q}{4\pi\varepsilon_o r}$

$F = QE$

$V = Ed$

$F = IlB\sin\theta$

$B = \frac{\mu_o I}{2\pi r}$

$c = \frac{1}{\sqrt{\varepsilon_o \mu_o}}$

$t = RC$

$X_C = \frac{V}{I}$

$X_C = \frac{1}{2\pi f C}$

$\varepsilon = -L\frac{dI}{dt}$

$E = \frac{1}{2}LI^2$

$X_L = \frac{V}{I}$

$X_L = 2\pi f L$

$\frac{\Delta W}{W} = \sqrt{\left(\frac{\Delta X}{X}\right)^2 + \left(\frac{\Delta Y}{Y}\right)^2 + \left(\frac{\Delta Z}{Z}\right)^2}$

$\Delta W = \sqrt{\Delta X^2 + \Delta Y^2 + \Delta Z^2}$

$d = \bar{v}t$

$s = \bar{v}t$

$v = u + at$

$s = ut + \frac{1}{2}at^2$

$v^2 = u^2 + 2as$

$s = \frac{1}{2}(u+v)t$

$W = mg$

$F = ma$

$E_W = Fd$

$E_P = mgh$

$E_K = \frac{1}{2}mv^2$

$P = \frac{E}{t}$

$p = mv$

$Ft = mv - mu$

$F = G\frac{Mm}{r^2}$

$t' = \dfrac{t}{\sqrt{1-\left(v/c\right)^2}}$

$l' = l\sqrt{1-\left(v/c\right)^2}$

$f_o = f_s\left(\dfrac{v}{v \pm v_s}\right)$

$z = \dfrac{\lambda_{observed} - \lambda_{rest}}{\lambda_{rest}}$

$z = \dfrac{v}{c}$

$v = H_0 d$

$W = QV$

$E = mc^2$

$E = hf$

$E_K = hf - hf_0$

$E_2 - E_1 = hf$

$T = \dfrac{1}{f}$

$v = f\lambda$

$d\sin\theta = m\lambda$

$n = \dfrac{\sin\theta_1}{\sin\theta_2}$

$\dfrac{\sin\theta_1}{\sin\theta_2} = \dfrac{\lambda_1}{\lambda_2} = \dfrac{v_1}{v_2}$

$\sin\theta_c = \dfrac{1}{n}$

$I = \dfrac{k}{d^2}$

$I = \dfrac{P}{A}$

path difference $= m\lambda$ or $\left(m+\dfrac{1}{2}\right)\lambda$ where $m = 0,1,2\ldots$

random uncertainty $= \dfrac{\text{max. value} - \text{min. value}}{\text{number of values}}$

$V_{peak} = \sqrt{2}V_{rms}$

$I_{peak} = \sqrt{2}I_{rms}$

$Q = It$

$V = IR$

$P = IV = I^2R = \dfrac{V^2}{R}$

$R_T = R_1 + R_2 + \ldots$

$\dfrac{1}{R_T} = \dfrac{1}{R_1} + \dfrac{1}{R_2} + \ldots$

$E = V + Ir$

$V_1 = \left(\dfrac{R_1}{R_1+R_2}\right)V_S$

$\dfrac{V_1}{V_2} = \dfrac{R_1}{R_2}$

$C = \dfrac{Q}{V}$

$E = \dfrac{1}{2}QV = \dfrac{1}{2}CV^2 = \dfrac{1}{2}\dfrac{Q^2}{C}$

Additional Relationships

Circle

circumference = $2\pi r$

area = πr^2

Sphere

area = $4\pi r^2$

volume = $\frac{4}{3}\pi r^3$

Trigonometry

$\sin\theta = \dfrac{\text{opposite}}{\text{hypotenuse}}$

$\cos\theta = \dfrac{\text{adjacent}}{\text{hypotenuse}}$

$\tan\theta = \dfrac{\text{opposite}}{\text{adjacent}}$

$\sin^2\theta + \cos^2\theta = 1$

Moment of inertia

point mass
$I = mr^2$

rod about centre
$I = \frac{1}{12}ml^2$

rod about end
$I = \frac{1}{3}ml^2$

disc about centre
$I = \frac{1}{2}mr^2$

sphere about centre
$I = \frac{2}{5}mr^2$

Table of standard derivatives

$f(x)$	$f'(x)$
$\sin ax$	$a\cos ax$
$\cos ax$	$-a\sin ax$

Table of standard integrals

$f(x)$	$\int f(x)dx$
$\sin ax$	$-\dfrac{1}{a}\cos ax + C$
$\cos ax$	$\dfrac{1}{a}\sin ax + C$

Electron Arrangements of Elements

Key

Atomic number
Symbol
Electron arrangement
Name

Main Groups

Group 1 (1)	Group 2 (2)		Group 3 (13)	Group 4 (14)	Group 5 (15)	Group 6 (16)	Group 7 (17)	Group 0 (18)
1 H 1 Hydrogen								2 He 2 Helium
3 Li 2,1 Lithium	4 Be 2,2 Beryllium		5 B 2,3 Boron	6 C 2,4 Carbon	7 N 2,5 Nitrogen	8 O 2,6 Oxygen	9 F 2,7 Fluorine	10 Ne 2,8 Neon
11 Na 2,8,1 Sodium	12 Mg 2,8,2 Magnesium		13 Al 2,8,3 Aluminium	14 Si 2,8,4 Silicon	15 P 2,8,5 Phosphorus	16 S 2,8,6 Sulphur	17 Cl 2,8,7 Chlorine	18 Ar 2,8,8 Argon

Transition Elements

(3)	(4)	(5)	(6)	(7)	(8)	(9)	(10)	(11)	(12)
21 Sc 2,8,9,2 Scandium	22 Ti 2,8,10,2 Titanium	23 V 2,8,11,2 Vanadium	24 Cr 2,8,13,1 Chromium	25 Mn 2,8,13,2 Manganese	26 Fe 2,8,14,2 Iron	27 Co 2,8,15,2 Cobalt	28 Ni 2,8,16,2 Nickel	29 Cu 2,8,18,1 Copper	30 Zn 2,8,18,2 Zinc
39 Y 2,8,18,9,2 Yttrium	40 Zr 2,8,18,10,2 Zirconium	41 Nb 2,8,18,12,1 Niobium	42 Mo 2,8,18,13,1 Molybdenum	43 Tc 2,8,18,13,2 Technetium	44 Ru 2,8,18,15,1 Ruthenium	45 Rh 2,8,18,16,1 Rhodium	46 Pd 2,8,18,18,0 Palladium	47 Ag 2,8,18,18,1 Silver	48 Cd 2,8,18,18,2 Cadmium
57 La 2,8,18,18,9,2 Lanthanum	72 Hf 2,8,18,32,10,2 Hafnium	73 Ta 2,8,18,32,11,2 Tantalum	74 W 2,8,18,32,12,2 Tungsten	75 Re 2,8,18,32,13,2 Rhenium	76 Os 2,8,18,32,14,2 Osmium	77 Ir 2,8,18,32,15,2 Iridium	78 Pt 2,8,18,32,17,1 Platinum	79 Au 2,8,18,32,18,1 Gold	80 Hg 2,8,18,32,18,2 Mercury
89 Ac 2,8,18,32,18,9,2 Actinium	104 Rf 2,8,18,32,32,10,2 Rutherfordium	105 Db 2,8,18,32,32,11,2 Dubnium	106 Sg 2,8,18,32,32,12,2 Seaborgium	107 Bh 2,8,18,32,32,13,2 Bohrium	108 Hs 2,8,18,32,32,14,2 Hassium	109 Mt 2,8,18,32,32,15,2 Meitnerium			

Groups 1–2 (continued) and 3–7 (continued)

Group 1	Group 2		Group 3	Group 4	Group 5	Group 6	Group 7	Group 0
19 K 2,8,8,1 Potassium	20 Ca 2,8,8,2 Calcium		31 Ga 2,8,18,3 Gallium	32 Ge 2,8,18,4 Germanium	33 As 2,8,18,5 Arsenic	34 Se 2,8,18,6 Selenium	35 Br 2,8,18,7 Bromine	36 Kr 2,8,18,8 Krypton
37 Rb 2,8,18,8,1 Rubidium	38 Sr 2,8,18,8,2 Strontium		49 In 2,8,18,18,3 Indium	50 Sn 2,8,18,18,4 Tin	51 Sb 2,8,18,18,5 Antimony	52 Te 2,8,18,18,6 Tellurium	53 I 2,8,18,18,7 Iodine	54 Xe 2,8,18,18,8 Xenon
55 Cs 2,8,18,18,8,1 Caesium	56 Ba 2,8,18,18,8,2 Barium		81 Tl 2,8,18,32,18,3 Thallium	82 Pb 2,8,18,32,18,4 Lead	83 Bi 2,8,18,32,18,5 Bismuth	84 Po 2,8,18,32,18,6 Polonium	85 At 2,8,18,32,18,7 Astatine	86 Rn 2,8,18,32,18,8 Radon
87 Fr 2,8,18,32,18,8,1 Francium	88 Ra 2,8,18,32,18,8,2 Radium							

Lanthanides

| 57 La 2,8,18,18,9,2 Lanthanum | 58 Ce 2,8,18,20,8,2 Cerium | 59 Pr 2,8,18,21,8,2 Praseodymium | 60 Nd 2,8,18,22,8,2 Neodymium | 61 Pm 2,8,18,23,8,2 Promethium | 62 Sm 2,8,18,24,8,2 Samarium | 63 Eu 2,8,18,25,8,2 Europium | 64 Gd 2,8,18,25,9,2 Gadolinium | 65 Tb 2,8,18,27,8,2 Terbium | 66 Dy 2,8,18,28,8,2 Dysprosium | 67 Ho 2,8,18,29,8,2 Holmium | 68 Er 2,8,18,30,8,2 Erbium | 69 Tm 2,8,18,31,8,2 Thulium | 70 Yb 2,8,18,32,8,2 Ytterbium | 71 Lu 2,8,18,32,9,2 Lutetium |

Actinides

| 89 Ac 2,8,18,32,18,9,2 Actinium | 90 Th 2,8,18,32,18,10,2 Thorium | 91 Pa 2,8,18,32,20,9,2 Protactinium | 92 U 2,8,18,32,21,9,2 Uranium | 93 Np 2,8,18,32,22,9,2 Neptunium | 94 Pu 2,8,18,32,24,8,2 Plutonium | 95 Am 2,8,18,32,25,8,2 Americium | 96 Cm 2,8,18,32,25,9,2 Curium | 97 Bk 2,8,18,32,27,8,2 Berkelium | 98 Cf 2,8,18,32,28,8,2 Californium | 99 Es 2,8,18,32,29,8,2 Einsteinium | 100 Fm 2,8,18,32,30,8,2 Fermium | 101 Md 2,8,18,32,31,8,2 Mendelevium | 102 No 2,8,18,32,32,8,2 Nobelium | 103 Lr 2,8,18,32,32,9,2 Lawrencium |

FOR OFFICIAL USE

National Qualifications 2016

Mark

X757/77/01

Physics

TUESDAY, 24 MAY
9:00 AM – 11:30 AM

Fill in these boxes and read what is printed below.

Full name of centre

Town

Forename(s)

Surname

Number of seat

Date of birth
Day   Month   Year   Scottish candidate number

Total marks — 140

Attempt ALL questions.

Reference may be made to the Physics Relationships Sheet X757/77/11 and the Data Sheet on *Page two*.

Write your answers clearly in the spaces provided in this booklet. Additional space for answers and rough work is provided at the end of this booklet. If you use this space you must clearly identify the question number you are attempting. Any rough work must be written in this booklet. You should score through your rough work when you have written your final copy.

Care should be taken to give an appropriate number of significant figures in the final answers to calculations.

Use **blue** or **black** ink.

Before leaving the examination room you must give this booklet to the Invigilator; if you do not, you may lose all the marks for this paper.

SQA

DATA SHEET
COMMON PHYSICAL QUANTITIES

Quantity	Symbol	Value	Quantity	Symbol	Value
Gravitational acceleration on Earth	g	9.8 m s^{-2}	Mass of electron	m_e	9.11×10^{-31} kg
Radius of Earth	R_E	6.4×10^6 m	Charge on electron	e	-1.60×10^{-19} C
Mass of Earth	M_E	6.0×10^{24} kg	Mass of neutron	m_n	1.675×10^{-27} kg
Mass of Moon	M_M	7.3×10^{22} kg	Mass of proton	m_p	1.673×10^{-27} kg
Radius of Moon	R_M	1.7×10^6 m	Mass of alpha particle	m_a	6.645×10^{-27} kg
Mean Radius of Moon Orbit		3.84×10^8 m	Charge on alpha particle		3.20×10^{-19} C
Solar radius		6.955×10^8 m	Planck's constant	h	6.63×10^{-34} J s
Mass of Sun		2.0×10^{30} kg	Permittivity of free space	ε_0	8.85×10^{-12} F m^{-1}
1 AU		1.5×10^{11} m	Permeability of free space	μ_0	$4\pi \times 10^{-7}$ H m^{-1}
Stefan-Boltzmann constant	σ	5.67×10^{-8} W m^{-2} K^{-4}	Speed of light in vacuum	c	3.0×10^8 m s^{-1}
Universal constant of gravitation	G	6.67×10^{-11} m^3 kg^{-1} s^{-2}	Speed of sound in air	v	3.4×10^2 m s^{-1}

REFRACTIVE INDICES
The refractive indices refer to sodium light of wavelength 589 nm and to substances at a temperature of 273 K.

Substance	Refractive index	Substance	Refractive index
Diamond	2.42	Glycerol	1.47
Glass	1.51	Water	1.33
Ice	1.31	Air	1.00
Perspex	1.49	Magnesium Fluoride	1.38

SPECTRAL LINES

Element	Wavelength/nm	Colour	Element	Wavelength/nm	Colour
Hydrogen	656	Red	Cadmium	644	Red
	486	Blue-green		509	Green
	434	Blue-violet		480	Blue
	410	Violet		Lasers	
	397	Ultraviolet	Element	Wavelength/nm	Colour
	389	Ultraviolet	Carbon dioxide	9550 } 10590 }	Infrared
Sodium	589	Yellow	Helium-neon	633	Red

PROPERTIES OF SELECTED MATERIALS

Substance	Density/ kg m^{-3}	Melting Point/ K	Boiling Point/K	Specific Heat Capacity/ J kg^{-1} K^{-1}	Specific Latent Heat of Fusion/ J kg^{-1}	Specific Latent Heat of Vaporisation/ J kg^{-1}
Aluminium	2.70×10^3	933	2623	9.02×10^2	3.95×10^5	....
Copper	8.96×10^3	1357	2853	3.86×10^2	2.05×10^5	....
Glass	2.60×10^3	1400	....	6.70×10^2	....	....
Ice	9.20×10^2	273	....	2.10×10^3	3.34×10^5	....
Glycerol	1.26×10^3	291	563	2.43×10^3	1.81×10^5	8.30×10^5
Methanol	7.91×10^2	175	338	2.52×10^3	9.9×10^4	1.12×10^6
Sea Water	1.02×10^3	264	377	3.93×10^3	....	....
Water	1.00×10^3	273	373	4.19×10^3	3.34×10^5	2.26×10^6
Air	1.29	....	....	....	....	....
Hydrogen	9.0×10^{-2}	14	20	1.43×10^4	....	4.50×10^5
Nitrogen	1.25	63	77	1.04×10^3	....	2.00×10^5
Oxygen	1.43	55	90	9.18×10^2	....	2.40×10^4

The gas densities refer to a temperature of 273 K and a pressure of 1.01×10^5 Pa.

Total marks — 140 marks

Attempt ALL questions

1.

   A car on a long straight track accelerates from rest. The car's run begins at time $t = 0$.

   Its velocity v at time t is given by the equation

   $$v = 0.135t^2 + 1.26t$$

   where v is measured in m s^{-1} and t is measured in s.

   Using **calculus** methods:

   (a) determine the acceleration of the car at $t = 15.0$ s;   3

   Space for working and answer

   (b) determine the displacement of the car from its original position at this time.   3

   Space for working and answer

2. (a) An ideal conical pendulum consists of a mass moving with constant speed in a circular path, as shown in Figure 2A.

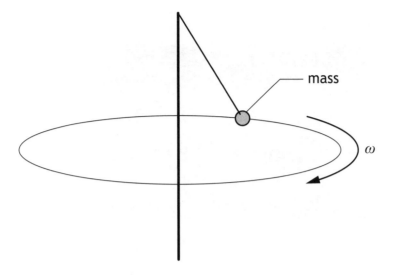

Figure 2A

(i) Explain why the mass is accelerating despite moving with constant speed. **1**

(ii) State the direction of this acceleration. **1**

2. (continued)

(b) Swingball is a garden game in which a ball is attached to a light string connected to a vertical pole as shown in Figure 2B.

The motion of the ball can be modelled as a conical pendulum.

The ball has a mass of 0·059 kg.

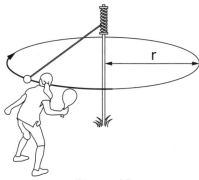

Figure 2B

(i) The ball is hit such that it moves with constant speed in a horizontal circle of radius 0·48 m.

The ball completes 1·5 revolutions in 2·69 s.

(A) Show that the angular velocity of the ball is 3·5 rad s^{-1}. 2

Space for working and answer

(B) Calculate the magnitude of the centripetal force acting on the ball. 3

Space for working and answer

2. (b) (i) (continued)

(C) The horizontal component of the tension in the string provides this centripetal force and the vertical component balances the weight of the ball.

Calculate the magnitude of the tension in the string. **3**

Space for working and answer

(ii) The string breaks whilst the ball is at the position shown in Figure 2C.

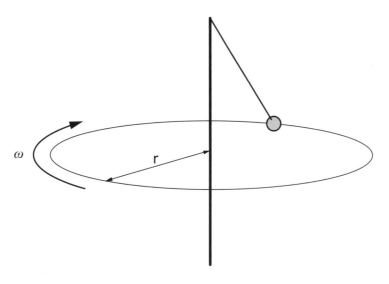

Figure 2C

On Figure 2C, draw the direction of the ball's velocity **immediately** after the string breaks. **1**

(An additional diagram, if required, can be found on *Page thirty-nine*.)

3. A spacecraft is orbiting a comet as shown in Figure 3.

   The comet can be considered as a sphere with a radius of $2{\cdot}1 \times 10^3$ m and a mass of $9{\cdot}5 \times 10^{12}$ kg.

   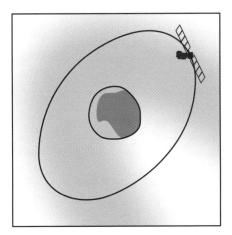

   Figure 3 (not to scale)

   (a) A lander was released by the spacecraft to land on the surface of the comet. After impact with the comet, the lander bounced back from the surface with an initial upward vertical velocity of $0{\cdot}38\,\text{m s}^{-1}$.

   By calculating the escape velocity of the comet, show that the lander returned to the surface for a second time. **4**

   Space for working and answer

3. (continued)

   (b) (i) Show that the gravitational field strength at the surface of the comet is $1.4 \times 10^{-4}\,N\,kg^{-1}$.

   Space for working and answer

   3

   (ii) Using the data from the space mission, a student tries to calculate the maximum height reached by the lander after its first bounce.

   The student's working is shown below

   $$v^2 = u^2 + 2as$$
   $$0 = 0.38^2 + 2 \times (-1.4 \times 10^{-4}) \times s$$
   $$s = 515.7\,m$$

   The actual maximum height reached by the lander was **not** as calculated by the student.

   State whether the actual maximum height reached would be greater or smaller than calculated by the student.

   You must justify your answer.

   3

4. Epsilon Eridani is a star 9.94×10^{16} m from Earth. It has a diameter of 1.02×10^9 m. The apparent brightness of Epsilon Eridani is measured on Earth to be 1.05×10^{-9} W m^{-2}.

(a) Calculate the luminosity of Epsilon Eridani. **3**

Space for working and answer

(b) Calculate the surface temperature of Epsilon Eridani. **3**

Space for working and answer

(c) State an assumption made in your calculation in (b). **1**

5. Einstein's theory of general relativity can be used to describe the motion of objects in non-inertial frames of reference. The equivalence principle is a key assumption of general relativity.

(a) Explain what is meant by the terms:

(i) *non-inertial frames of reference*;

(ii) *the equivalence principle*.

(b) Two astronauts are on board a spacecraft in deep space far away from any large masses. When the spacecraft is accelerating one astronaut throws a ball towards the other.

(i) On Figure 5A sketch the path that the ball would follow in the astronauts' frame of reference.

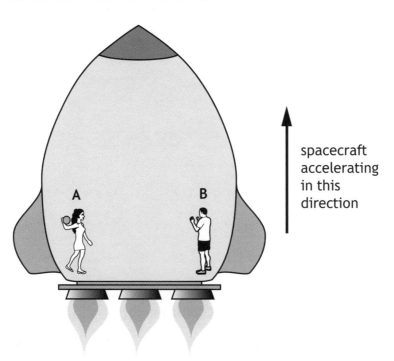

Figure 5A

(An additional diagram, if required, can be found on *Page thirty-nine*.)

5. (b) (continued)

   (ii) The experiment is repeated when the spacecraft is travelling at constant speed.

   On Figure 5B sketch the path that the ball would follow in the astronauts' frame of reference. **1**

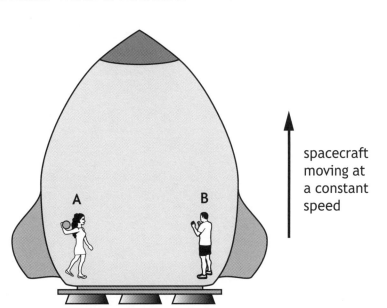

   Figure 5B

   (An additional diagram, if required, can be found on *Page forty*.)

(c) A clock is on the surface of the Earth and an identical clock is on board a spacecraft which is accelerating in deep space at $8\,\text{m s}^{-2}$.

   State which clock runs slower.

   Justify your answer in terms of the equivalence principle. **2**

[Turn over

6. A student makes the following statement.

"Quantum theory — I don't understand it. I don't really know what it is. I believe that classical physics can explain everything."

Use your knowledge of physics to comment on the statement. **3**

7. (a) The Earth can be modelled as a black body radiator.

The average surface temperature of the Earth can be estimated using the relationship

$$T = \frac{b}{\lambda_{peak}}$$

where

T is the average surface temperature of the Earth in kelvin;

b is Wien's Displacement Constant equal to $2 \cdot 89 \times 10^{-3}$ K m;

λ_{peak} is the peak wavelength of the radiation emitted by a black body radiator.

The average surface temperature of Earth is 15 °C.

(i) Estimate the peak wavelength of the radiation emitted by Earth. **3**

Space for working and answer

(ii) To which part of the electromagnetic spectrum does this peak wavelength correspond? **1**

[Turn over

7. (continued)

(b) In order to investigate the properties of black body radiators a student makes measurements from the spectra produced by a filament lamp. Measurements are made when the lamp is operated at its rated voltage and when it is operated at a lower voltage.

The filament lamp can be considered to be a black body radiator.

A graph of the results obtained is shown in Figure 7.

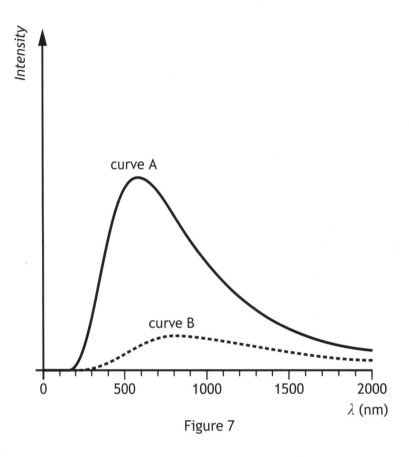

Figure 7

(i) State which curve corresponds to the radiation emitted when the filament lamp is operating at its rated voltage.

You must justify your answer. **2**

(ii) The shape of the curves on the graph on Figure 7 is not as predicted by classical physics.

On Figure 7, sketch a curve to show the result predicted by classical physics. **1**

(An additional graph, if required, can be found on *Page forty*.)

8. Werner Heisenberg is considered to be one of the pioneers of quantum mechanics.

   He is most famous for his uncertainty principle which can be expressed in the equation

   $$\Delta x \Delta p_x \geq \frac{h}{4\pi}$$

   (a) (i) State what quantity is represented by the term Δp_x. 1

   (ii) Explain the implications of the Heisenberg uncertainty principle for experimental measurements. 1

[Turn over

8. (continued)

(b) In an experiment to investigate the nature of particles, individual electrons were fired one at a time from an electron gun through a narrow double slit. The position where each electron struck the detector was recorded and displayed on a computer screen.

The experiment continued until a clear pattern emerged on the screen as shown in Figure 8.

The momentum of each electron at the double slit is 6.5×10^{-24} kg m s^{-1}.

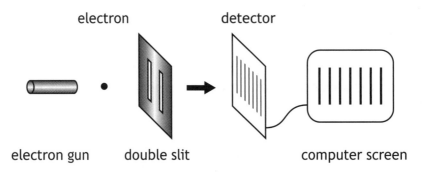

Figure 8   not to scale

(i) The experimenter had three different double slits with slit separations 0·1 mm, 0·1 μm and 0·1 nm.

State which double slit was used to produce the image on the screen.

You must justify your answer by calculation of the de Broglie wavelength.

4

Space for working and answer

8. (b) (continued)

(ii) The uncertainty in the momentum of an electron at the double slit is 6.5×10^{-26} kg m s^{-1}.

Calculate the minimum absolute uncertainty in the position of the electron.

Space for working and answer

(iii) Explain fully how the experimental result shown in Figure 8 can be interpreted.

[Turn over

9. A particle with charge q and mass m is travelling with constant speed v. The particle enters a uniform magnetic field at 90° and is forced to move in a circle of radius r as shown in Figure 9.

The magnetic induction of the field is B.

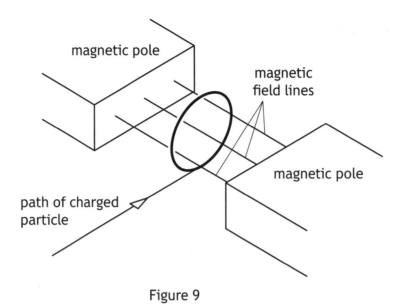

Figure 9

(a) Show that the radius of the circular path of the particle is given by

$$r = \frac{mv}{Bq}$$

2

9. **(continued)**

(b) In an experimental nuclear reactor, charged particles are contained in a magnetic field. One such particle is a deuteron consisting of one proton and one neutron.

The kinetic energy of each deuteron is 1·50 MeV.

The mass of the deuteron is $3·34 \times 10^{-27}$ kg.

Relativistic effects can be ignored.

(i) Calculate the speed of the deuteron. **4**

Space for working and answer

(ii) Calculate the magnetic induction required to keep the deuteron moving in a circular path of radius 2·50 m. **2**

Space for working and answer

[Turn over

9. (b) (continued)

(iii) Deuterons are fused together in the reactor to produce isotopes of helium.

$_{2}^{3}$He nuclei, each comprising 2 protons and 1 neutron, are present in the reactor.

A $_{2}^{3}$He nucleus also moves in a circular path in the same magnetic field.

The $_{2}^{3}$He nucleus moves at the same speed as the deuteron.

State whether the radius of the circular path of the $_{2}^{3}$He nucleus is greater than, equal to or less than 2·50 m.

You must justify your answer.   **2**

10. (a) (i) State what is meant by *simple harmonic motion*.

(ii) The displacement of an oscillating object can be described by the expression

$$y = A\cos\omega t$$

where the symbols have their usual meaning.

Show that this expression is a solution to the equation

$$\frac{d^2y}{dt^2} + \omega^2 y = 0$$

[Turn over

10. (continued)

(b) A mass of 1·5 kg is suspended from a spring of negligible mass as shown in Figure 10. The mass is displaced downwards 0·040 m from its equilibrium position.

The mass is then released from this position and begins to oscillate. The mass completes ten oscillations in a time of 12 s.

Frictional forces can be considered to be negligible.

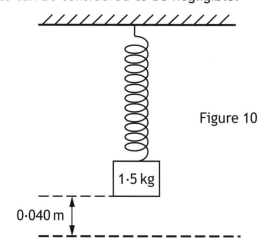

Figure 10

(i) Show that the angular frequency ω of the mass is 5·2 rad s^{-1}.   **3**

Space for working and answer

(ii) Calculate the maximum velocity of the mass.   **3**

Space for working and answer

10. (b) (continued)

(iii) Determine the potential energy stored in the spring when the mass is at its maximum displacement. **3**

Space for working and answer

(c) The system is now modified so that a damping force acts on the oscillating mass.

(i) Describe how this modification may be achieved. **1**

(ii) Using the axes below sketch a graph showing, for the modified system, how the displacement of the mass varies with time after release.

Numerical values are **not** required on the axes. **1**

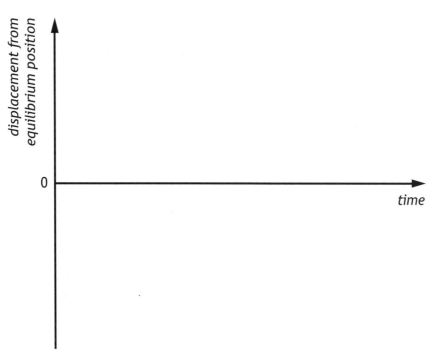

(An additional graph, if required, can be found on *Page forty-one.*)

11.

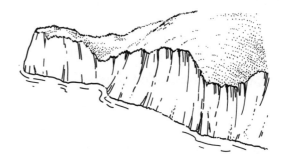

foghorn

A ship emits a blast of sound from its foghorn. The sound wave is described by the equation

$$y = 0\cdot 250 \sin 2\pi (118t - 0\cdot 357x)$$

where the symbols have their usual meaning.

(a) Determine the speed of the sound wave.

Space for working and answer

11. (continued)

(b) The sound from the ship's foghorn reflects from a cliff. When it reaches the ship this reflected sound has half the energy of the original sound.

Write an equation describing the reflected sound wave at this point.

4

[Turn over

12. Some early 3D video cameras recorded two separate images at the same time to create two almost identical movies.

Cinemas showed 3D films by projecting these two images simultaneously onto the same screen using two projectors. Each projector had a polarising filter through which the light passed as shown in Figure 12.

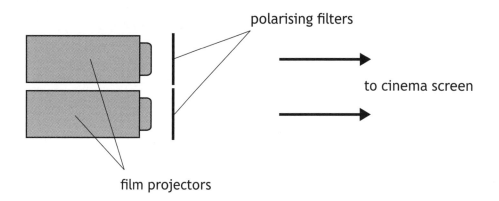

Figure 12

(a) Describe how the transmission axes of the two polarising filters should be arranged so that the two images on the screen do not interfere with each other. **1**

(b) A student watches a 3D movie using a pair of glasses which contains two polarising filters, one for each eye.

Explain how this arrangement enables a different image to be seen by each eye. **2**

12. (continued)

(c) Before the film starts, the student looks at a ceiling lamp through one of the filters in the glasses. While looking at the lamp, the student then rotates the filter through 90°.

State what effect, if any, this rotation will have on the observed brightness of the lamp.

Justify your answer. **2**

(d) During the film, the student looks at the screen through only one of the filters in the glasses. The student then rotates the filter through 90° and does not observe any change in brightness.

Explain this observation. **1**

[Turn over

13. (a) Q_1 is a point charge of +12 nC. Point Y is 0·30 m from Q_1 as shown in Figure 13A.

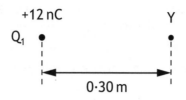

Figure 13A

Show that the electrical potential at point Y is +360 V. **2**

Space for working and answer

(b) A second point charge Q_2 is placed at a distance of 0·40 m from point Y as shown in Figure 13B. The electrical potential at point Y is now zero.

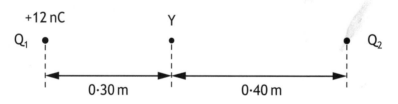

Figure 13B

(i) Determine the charge of Q_2. **3**

Space for working and answer

13. (b) (continued)

    (ii) Determine the electric field strength at point Y.   **4**

        Space for working and answer

    (iii) On Figure 13C, sketch the electric field pattern for this system of charges.   **2**

Q_1 •   • Q_2

Figure 13C

(An additional diagram, if required, can be found on *Page forty-one*.)

[Turn over

14. A student measures the magnetic induction at a distance r from a long straight current carrying wire using the apparatus shown in Figure 14.

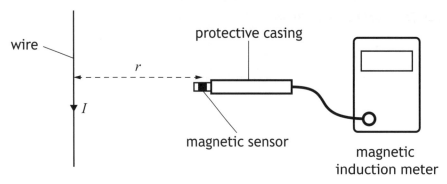

Figure 14

The following data are obtained.

Distance from wire $r = 0\cdot10\,\text{m}$
Magnetic induction $B = 5\cdot0\,\mu\text{T}$

(a) Use the data to calculate the current I in the wire.

Space for working and answer

(b) The student estimates the following uncertainties in the measurements of B and r.

Uncertainties in r		Uncertainties in B	
reading	±0·002 m	reading	±0·1 µT
calibration	±0·0005 m	calibration	±1·5% of reading

(i) Calculate the percentage uncertainty in the measurement of r.

Space for working and answer

14. (b) (continued)

(ii) Calculate the percentage uncertainty in the measurement of B. **3**

Space for working and answer

(iii) Calculate the absolute uncertainty in the value of the current in the wire. **2**

Space for working and answer

(c) The student measures distance r, as shown in Figure 14, using a metre stick. The smallest scale division on the metre stick is 1 mm.

Suggest a reason why the student's estimate of the reading uncertainty in r is not ± 0.5 mm. **1**

[Turn over

15. A student constructs a simple air-insulated capacitor using two parallel metal plates, each of area A, separated by a distance d. The plates are separated using small insulating spacers as shown in Figure 15A.

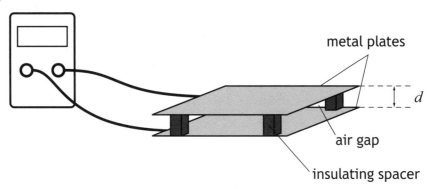

Figure 15A

The capacitance C of the capacitor is given by

$$C = \varepsilon_0 \frac{A}{d}$$

The student investigates how the capacitance depends on the separation of the plates. The student uses a capacitance meter to measure the capacitance for different plate separations. The plate separation is measured using a ruler.

The results are used to plot the graph shown in Figure 15B.

The area of each metal plate is 9.0×10^{-2} m².

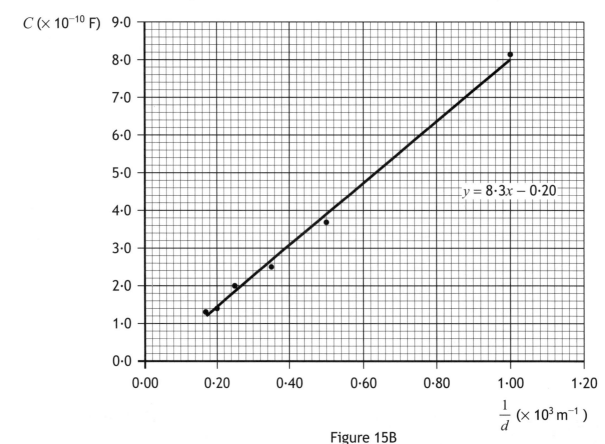

Figure 15B

15. (continued)

(a) (i) Use information from the graph to determine a value for ε_o, the permittivity of free space. **3**

Space for working and answer

(ii) Use your calculated value for the permittivity of free space to determine a value for the speed of light in air. **3**

Space for working and answer

(b) The best fit line on the graph does not pass through the origin as theory predicts.

Suggest a reason for this. **1**

[Turn over

16. A student uses two methods to determine the moment of inertia of a solid sphere about an axis through its centre.

(a) In the first method the student measures the mass of the sphere to be 3·8 kg and the radius to be 0·053 m.

Calculate the moment of inertia of the sphere. **3**

Space for working and answer

(b) In the second method, the student uses conservation of energy to determine the moment of inertia of the sphere.

The following equation describes the conservation of energy as the sphere rolls down the slope

$$mgh = \frac{1}{2}mv^2 + \frac{1}{2}I\omega^2$$

where the symbols have their usual meanings.

The equation can be rearranged to give the following expression

$$2gh = \left(\frac{I}{mr^2} + 1\right)v^2$$

This expression is in the form of the equation of a straight line through the origin,

$$y = gradient \times x$$

[Turn over

16. (b) (continued)

The student measures the height of the slope h. The student then allows the sphere to roll down the slope and measures the final speed of the sphere v at the bottom of the slope as shown in Figure 16.

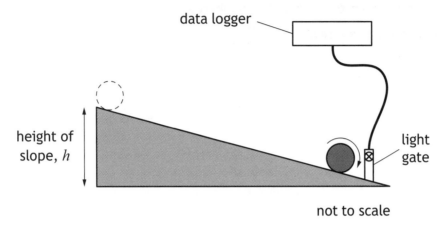

Figure 16

The following is an extract from the student's notebook.

h (m)	v (m s^{-1})	$2gh$ (m^2 s^{-2})	v^2 (m^2 s^{-2})
0·020	0·42	0·39	0·18
0·040	0·63	0·78	0·40
0·060	0·68	1·18	0·46
0·080	0·95	1·57	0·90
0·100	1·05	1·96	1·10

$m = 3·8$ kg $r = 0·053$ m

(i) On the square-ruled paper on *Page thirty-seven*, draw a graph that would allow the student to determine the moment of inertia of the sphere. **3**

(ii) Use the gradient of your line to determine the moment of inertia of the sphere. **3**

Space for working and answer

(An additional square-ruled paper, if required, can be found on *Page forty-two*.)

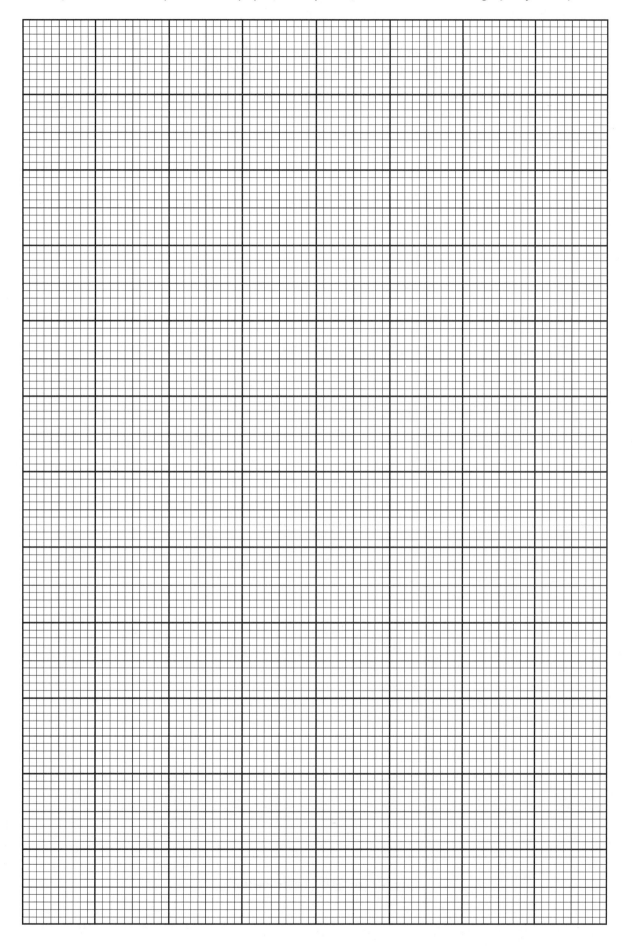

[Turn over for next question

16. (continued)

(c) The student states that more confidence should be placed in the value obtained for the moment of inertia in the second method.

Use your knowledge of experimental physics to comment on the student's statement.

3

[END OF QUESTION PAPER]

ADDITIONAL SPACE FOR ANSWERS AND ROUGH WORK

Additional diagram for Question 2 (b) (ii)

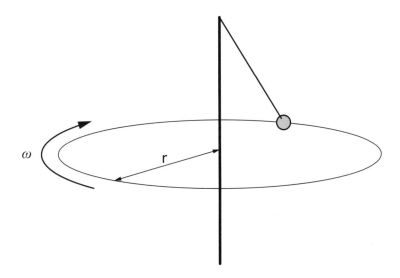

Figure 2C

Additional diagram for Question 5 (b) (i)

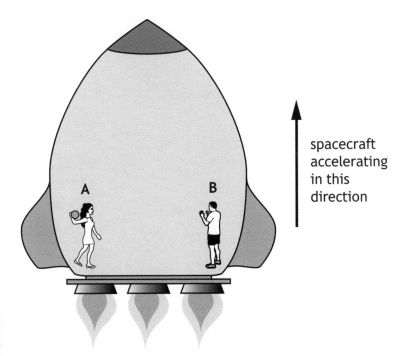

Figure 5A

ADDITIONAL SPACE FOR ANSWERS AND ROUGH WORK

Additional diagram for Question 5 (b) (ii)

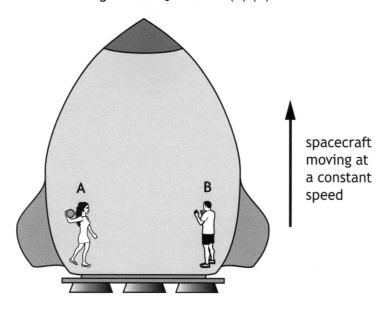

Figure 5B

Additional diagram for Question 7 (b) (ii)

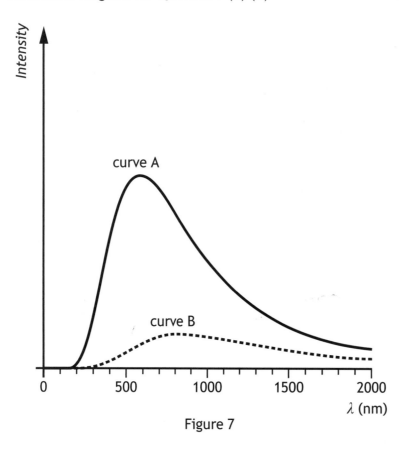

Figure 7

ADDITIONAL SPACE FOR ANSWERS AND ROUGH WORK

Additional diagram for Question 10 (c) (ii)

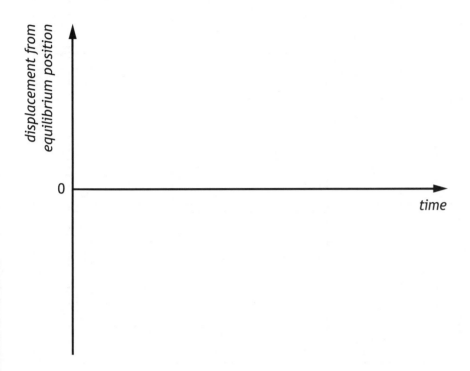

Additional diagram for Question 13 (b) (iii)

Q₁ •   • Q₂

Figure 13C

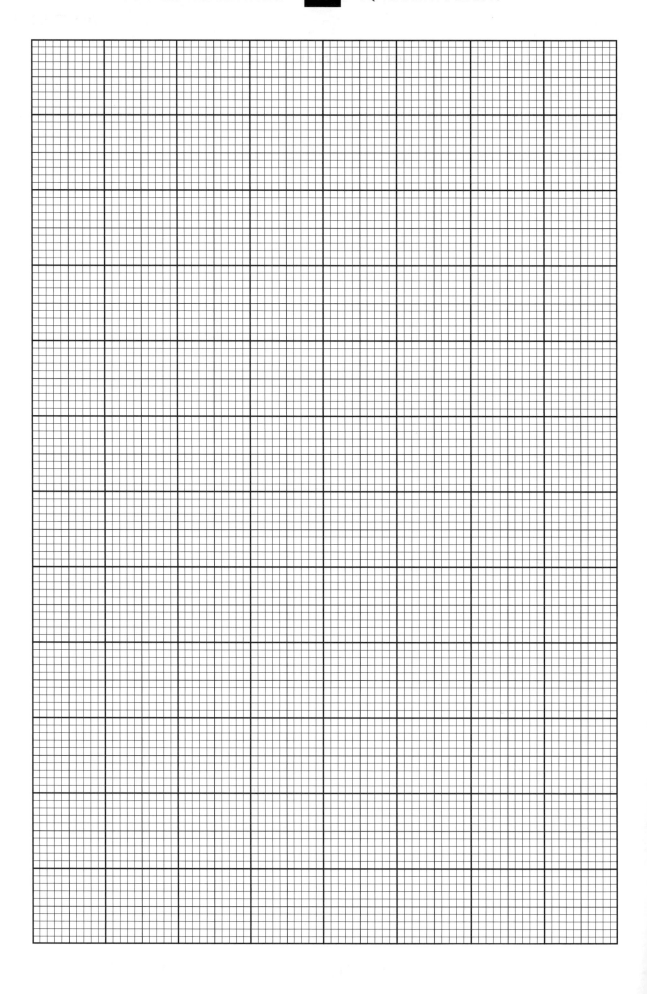

ADDITIONAL SPACE FOR ANSWERS AND ROUGH WORK

ADDITIONAL SPACE FOR ANSWERS AND ROUGH WORK

ADVANCED HIGHER
2017

X757/77/11

Physics
Relationships Sheet

National Qualifications 2017

WEDNESDAY, 17 MAY
9:00 AM – 11:30 AM

Relationships required for Physics Advanced Higher

$v = \dfrac{ds}{dt}$

$a = \dfrac{dv}{dt} = \dfrac{d^2s}{dt^2}$

$v = u + at$

$s = ut + \dfrac{1}{2}at^2$

$v^2 = u^2 + 2as$

$\omega = \dfrac{d\theta}{dt}$

$\alpha = \dfrac{d\omega}{dt} = \dfrac{d^2\theta}{dt^2}$

$\omega = \omega_o + \alpha t$

$\theta = \omega_o t + \dfrac{1}{2}\alpha t^2$

$\omega^2 = \omega_o^2 + 2\alpha\theta$

$s = r\theta$

$v = r\omega$

$a_t = r\alpha$

$a_r = \dfrac{v^2}{r} = r\omega^2$

$F = \dfrac{mv^2}{r} = mr\omega^2$

$T = Fr$

$T = I\alpha$

$L = mvr = mr^2\omega$

$L = I\omega$

$E_K = \dfrac{1}{2}I\omega^2$

$F = G\dfrac{Mm}{r^2}$

$V = -\dfrac{GM}{r}$

$v = \sqrt{\dfrac{2GM}{r}}$

apparent brightness, $b = \dfrac{L}{4\pi r^2}$

Power per unit area $= \sigma T^4$

$L = 4\pi r^2 \sigma T^4$

$r_{Schwarzschild} = \dfrac{2GM}{c^2}$

$E = hf$

$\lambda = \dfrac{h}{p}$

$mvr = \dfrac{nh}{2\pi}$

$\Delta x\, \Delta p_x \geq \dfrac{h}{4\pi}$

$\Delta E\, \Delta t \geq \dfrac{h}{4\pi}$

$F = qvB$

$\omega = 2\pi f$

$a = \dfrac{d^2y}{dt^2} = -\omega^2 y$

$y = A\cos\omega t$ or $y = A\sin\omega t$

$v = \pm\omega\sqrt{(A^2 - y^2)}$

$E_K = \frac{1}{2}m\omega^2(A^2 - y^2)$

$E_P = \frac{1}{2}m\omega^2 y^2$

$y = A\sin 2\pi(ft - \frac{x}{\lambda})$

$E = kA^2$

$\phi = \frac{2\pi x}{\lambda}$

optical path difference $= m\lambda$ or $\left(m+\frac{1}{2}\right)\lambda$

where $m = 0, 1, 2...$

$\Delta x = \frac{\lambda l}{2d}$

$d = \frac{\lambda}{4n}$

$\Delta x = \frac{\lambda D}{d}$

$n = \tan i_p$

$F = \frac{Q_1 Q_2}{4\pi\varepsilon_o r^2}$

$E = \frac{Q}{4\pi\varepsilon_o r^2}$

$V = \frac{Q}{4\pi\varepsilon_o r}$

$F = QE$

$V = Ed$

$F = IlB\sin\theta$

$B = \frac{\mu_o I}{2\pi r}$

$c = \frac{1}{\sqrt{\varepsilon_o \mu_o}}$

$t = RC$

$X_C = \frac{V}{I}$

$X_C = \frac{1}{2\pi f C}$

$\varepsilon = -L\frac{dI}{dt}$

$E = \frac{1}{2}LI^2$

$X_L = \frac{V}{I}$

$X_L = 2\pi f L$

$\frac{\Delta W}{W} = \sqrt{\left(\frac{\Delta X}{X}\right)^2 + \left(\frac{\Delta Y}{Y}\right)^2 + \left(\frac{\Delta Z}{Z}\right)^2}$

$\Delta W = \sqrt{\Delta X^2 + \Delta Y^2 + \Delta Z^2}$

$$d = \bar{v}t$$

$$s = \bar{v}t$$

$$v = u + at$$

$$s = ut + \frac{1}{2}at^2$$

$$v^2 = u^2 + 2as$$

$$s = \frac{1}{2}(u+v)t$$

$$W = mg$$

$$F = ma$$

$$E_W = Fd$$

$$E_P = mgh$$

$$E_K = \frac{1}{2}mv^2$$

$$P = \frac{E}{t}$$

$$p = mv$$

$$Ft = mv - mu$$

$$F = G\frac{Mm}{r^2}$$

$$t' = \frac{t}{\sqrt{1-\left(v/c\right)^2}}$$

$$l' = l\sqrt{1-\left(v/c\right)^2}$$

$$f_o = f_s\left(\frac{v}{v \pm v_s}\right)$$

$$z = \frac{\lambda_{observed} - \lambda_{rest}}{\lambda_{rest}}$$

$$z = \frac{v}{c}$$

$$v = H_0 d$$

$$W = QV$$

$$E = mc^2$$

$$E = hf$$

$$E_K = hf - hf_0$$

$$E_2 - E_1 = hf$$

$$T = \frac{1}{f}$$

$$v = f\lambda$$

$$d\sin\theta = m\lambda$$

$$n = \frac{\sin\theta_1}{\sin\theta_2}$$

$$\frac{\sin\theta_1}{\sin\theta_2} = \frac{\lambda_1}{\lambda_2} = \frac{v_1}{v_2}$$

$$\sin\theta_c = \frac{1}{n}$$

$$I = \frac{k}{d^2}$$

$$I = \frac{P}{A}$$

path difference $= m\lambda$ or $\left(m+\frac{1}{2}\right)\lambda$ where $m = 0, 1, 2...$

random uncertainty $= \dfrac{\text{max. value} - \text{min. value}}{\text{number of values}}$

$$V_{peak} = \sqrt{2}V_{rms}$$

$$I_{peak} = \sqrt{2}I_{rms}$$

$$Q = It$$

$$V = IR$$

$$P = IV = I^2R = \frac{V^2}{R}$$

$$R_T = R_1 + R_2 + ...$$

$$\frac{1}{R_T} = \frac{1}{R_1} + \frac{1}{R_2} + ...$$

$$E = V + Ir$$

$$V_1 = \left(\frac{R_1}{R_1 + R_2}\right)V_S$$

$$\frac{V_1}{V_2} = \frac{R_1}{R_2}$$

$$C = \frac{Q}{V}$$

$$E = \frac{1}{2}QV = \frac{1}{2}CV^2 = \frac{1}{2}\frac{Q^2}{C}$$

Additional Relationships

Circle

circumference $= 2\pi r$

area $= \pi r^2$

Sphere

area $= 4\pi r^2$

volume $= \frac{4}{3}\pi r^3$

Trigonometry

$\sin\theta = \dfrac{\text{opposite}}{\text{hypotenuse}}$

$\cos\theta = \dfrac{\text{adjacent}}{\text{hypotenuse}}$

$\tan\theta = \dfrac{\text{opposite}}{\text{adjacent}}$

$\sin^2\theta + \cos^2\theta = 1$

Moment of inertia

point mass
$I = mr^2$

rod about centre
$I = \frac{1}{12}ml^2$

rod about end
$I = \frac{1}{3}ml^2$

disc about centre
$I = \frac{1}{2}mr^2$

sphere about centre
$I = \frac{2}{5}mr^2$

Table of standard derivatives

$f(x)$	$f'(x)$
$\sin ax$	$a\cos ax$
$\cos ax$	$-a\sin ax$

Table of standard integrals

$f(x)$	$\int f(x)dx$
$\sin ax$	$-\dfrac{1}{a}\cos ax + C$
$\cos ax$	$\dfrac{1}{a}\sin ax + C$

AH

National Qualifications 2017

X757/77/01

Physics

WEDNESDAY, 17 MAY
9:00 AM – 11:30 AM

Fill in these boxes and read what is printed below.

Full name of centre

Town

Forename(s)

Surname

Number of seat

Date of birth
Day Month Year Scottish candidate number

Total marks — 140

Attempt ALL questions.

Reference may be made to the Physics Relationship Sheet X757/77/11 and the Data Sheet on *Page two*.

Write your answers clearly in the spaces provided in this booklet. Additional space for answers and rough work is provided at the end of this booklet. If you use this space you must clearly identify the question number you are attempting. Any rough work must be written in this booklet. You should score through your rough work when you have written your final copy.

Care should be taken to give an appropriate number of significant figures in the final answers to calculations.

Use **blue** or **black** ink.

Before leaving the examination room you must give this booklet to the Invigilator; if you do not, you may lose all the marks for this paper.

SQA

DATA SHEET

COMMON PHYSICAL QUANTITIES

Quantity	Symbol	Value	Quantity	Symbol	Value
Gravitational acceleration on Earth	g	9.8 m s^{-2}	Mass of electron	m_e	9.11×10^{-31} kg
Radius of Earth	R_E	6.4×10^6 m	Charge on electron	e	-1.60×10^{-19} C
Mass of Earth	M_E	6.0×10^{24} kg	Mass of neutron	m_n	1.675×10^{-27} kg
Mass of Moon	M_M	7.3×10^{22} kg	Mass of proton	m_p	1.673×10^{-27} kg
Radius of Moon	R_M	1.7×10^6 m	Mass of alpha particle	m_α	6.645×10^{-27} kg
Mean Radius of Moon Orbit		3.84×10^8 m	Charge on alpha particle		3.20×10^{-19} C
Solar radius		6.955×10^8 m	Planck's constant	h	6.63×10^{-34} J s
Mass of Sun		2.0×10^{30} kg	Permittivity of free space	ε_0	8.85×10^{-12} F m^{-1}
1 AU		1.5×10^{11} m			
Stefan-Boltzmann constant	σ	5.67×10^{-8} W m^{-2} K^{-4}	Permeability of free space	μ_0	$4\pi \times 10^{-7}$ H m^{-1}
Universal constant of gravitation	G	6.67×10^{-11} m^3 kg^{-1} s^{-2}	Speed of light in vacuum	c	3.00×10^8 m s^{-1}
			Speed of sound in air	v	3.4×10^2 m s^{-1}

REFRACTIVE INDICES

The refractive indices refer to sodium light of wavelength 589 nm and to substances at a temperature of 273 K.

Substance	Refractive index	Substance	Refractive index
Diamond	2·42	Glycerol	1·47
Glass	1·51	Water	1·33
Ice	1·31	Air	1·00
Perspex	1·49	Magnesium Fluoride	1·38

SPECTRAL LINES

Element	Wavelength/nm	Colour	Element	Wavelength/nm	Colour
Hydrogen	656	Red	Cadmium	644	Red
	486	Blue-green		509	Green
	434	Blue-violet		480	Blue
	410	Violet	Lasers		
	397	Ultraviolet	Element	Wavelength/nm	Colour
	389	Ultraviolet	Carbon dioxide	9550 } 10590 }	Infrared
Sodium	589	Yellow	Helium-neon	633	Red

PROPERTIES OF SELECTED MATERIALS

Substance	Density/ kg m^{-3}	Melting Point/ K	Boiling Point/K	Specific Heat Capacity/ J kg^{-1} K^{-1}	Specific Latent Heat of Fusion/ J kg^{-1}	Specific Latent Heat of Vaporisation/ J kg^{-1}
Aluminium	2.70×10^3	933	2623	9.02×10^2	3.95×10^5	....
Copper	8.96×10^3	1357	2853	3.86×10^2	2.05×10^5	....
Glass	2.60×10^3	1400	....	6.70×10^2	....	....
Ice	9.20×10^2	273	....	2.10×10^3	3.34×10^5	....
Glycerol	1.26×10^3	291	563	2.43×10^3	1.81×10^5	8.30×10^5
Methanol	7.91×10^2	175	338	2.52×10^3	9.9×10^4	1.12×10^6
Sea Water	1.02×10^3	264	377	3.93×10^3	....	....
Water	1.00×10^3	273	373	4.18×10^3	3.34×10^5	2.26×10^6
Air	1·29	....	....	....	....	....
Hydrogen	9.0×10^{-2}	14	20	1.43×10^4	....	4.50×10^5
Nitrogen	1·25	63	77	1.04×10^3	....	2.00×10^5
Oxygen	1·43	55	90	9.18×10^2	....	2.40×10^4

The gas densities refer to a temperature of 273 K and a pressure of 1.01×10^5 Pa.

Total marks — 140 marks

Attempt ALL questions

1. An athlete competes in a one hundred metre race on a flat track, as shown in Figure 1A.

Figure 1A

Starting from rest, the athlete's speed for the first 3·10 seconds of the race can be modelled using the relationship

$$v = 0\cdot 4t^2 + 2t$$

where the symbols have their usual meaning.

According to this model:

(a) determine the speed of the athlete at $t = 3\cdot 10\,\text{s}$; 2

Space for working and answer

(b) determine, using **calculus** methods, the distance travelled by the athlete in this time. 3

Space for working and answer

[Turn over

2. (a) As part of a lesson, a teacher swings a sphere tied to a light string as shown in Figure 2A. The path of the sphere is a vertical circle as shown in Figure 2B.

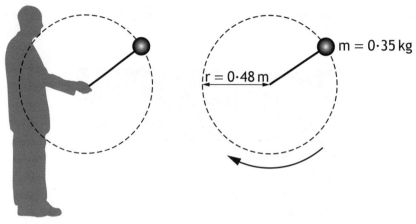

Figure 2A   Figure 2B

(i) On Figure 2C, show the forces acting on the sphere as it passes through its highest point.

You must name these forces and show their directions. 1

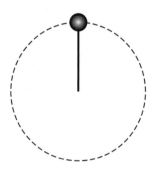

Figure 2C

MARKS DO NOT WRITE IN THIS MARGIN

2. (a) (continued)

(ii) On Figure 2D, show the forces acting on the sphere as it passes through its lowest point.

You must name these forces and show their directions. 1

Figure 2D

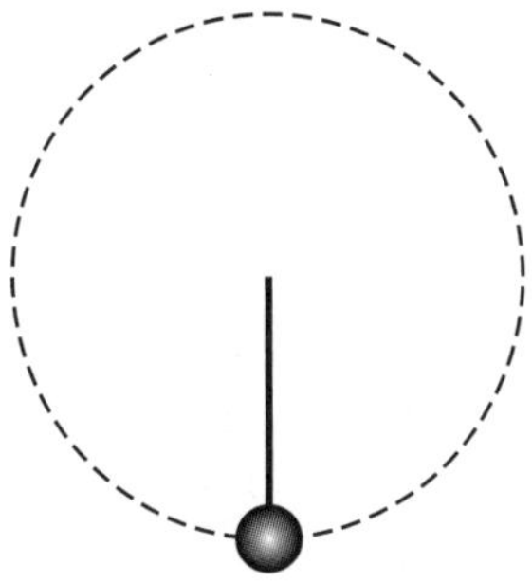

(iii) The sphere of mass 0·35 kg can be considered to be moving at a constant speed.

The centripetal force acting on the sphere is 4·0 N.

Determine the magnitude of the tension in the light string when the sphere is at:

(A) the highest position in its circular path; 2

Space for working and answer

(B) the lowest position in its circular path. 1

Space for working and answer

2. (continued)

(b) The speed of the sphere is now gradually reduced until the sphere no longer travels in a circular path.

Explain why the sphere no longer travels in a circular path. 2

(c) The teacher again swings the sphere with constant speed in a vertical circle. The student shown in Figure 2E observes the sphere moving up and down vertically with simple harmonic motion.

The period of this motion is 1·4 s.

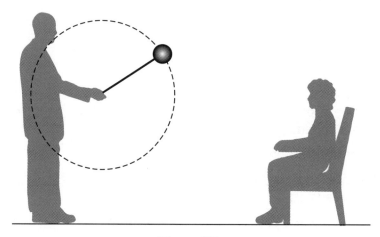

Figure 2E

Figure 2F represents the path of the sphere as observed by the student.

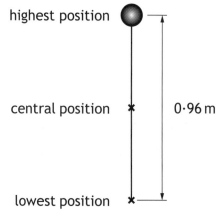

Figure 2F

2. (c) (continued)

On Figure 2G, sketch a graph showing how the vertical displacement *s* of the sphere from its central position varies with time *t*, as it moves from its highest position to its lowest position.

Numerical values are required on both axes. 3

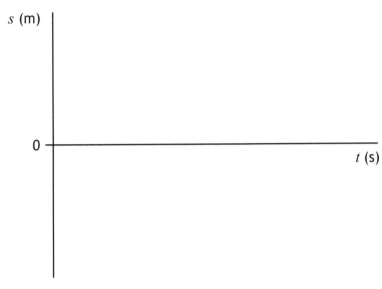

Figure 2G

(An additional diagram, if required, can be found on *Page forty-two*.)

[Turn over

3. A student uses a solid, uniform circular disc of radius 290 mm and mass 0·40 kg as part of an investigation into rotational motion.

The disc is shown in Figure 3A.

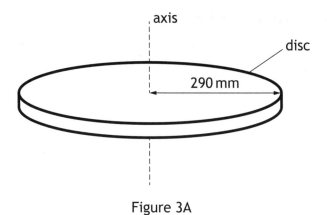

Figure 3A

(a) Calculate the moment of inertia of the disc about the axis shown in Figure 3A.

Space for working and answer

3. (continued)

(b) The disc is now mounted horizontally on the axle of a rotational motion sensor as shown in Figure 3B.

The axle is on a frictionless bearing. A thin cord is wound around a stationary pulley which is attached to the axle.

The moment of inertia of the pulley and axle can be considered negligible.

The pulley has a radius of 7·5 mm and a force of 8·0 N is applied to the free end of the cord.

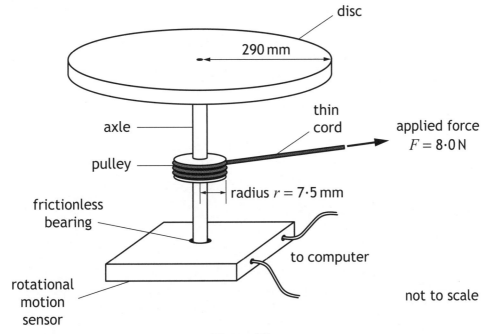

Figure 3B

(i) Calculate the torque applied to the pulley.

Space for working and answer

(ii) Calculate the angular acceleration produced by this torque.

Space for working and answer

3. (b) (continued)

(iii) The cord becomes detached from the pulley after 0·25 m has unwound.

By considering the angular displacement of the disc, determine its angular velocity when the cord becomes detached. **5**

Space for working and answer

3. (continued)

(c) In a second experiment the disc has an angular velocity of 12 rad s^{-1}.

The student now drops a small 25 g cube vertically onto the disc. The cube sticks to the disc.

The centre of mass of the cube is 220 mm from the axis of rotation, as shown in Figure 3C.

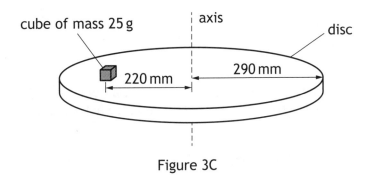

Figure 3C

Calculate the angular velocity of the system immediately after the cube was dropped onto the disc.

Space for working and answer

4. The NASA space probe Dawn has travelled to and orbited large asteroids in the solar system. Dawn has a mass of 1240 kg.

   The table gives information about two large asteroids orbited by Dawn. Both asteroids can be considered to be spherical and remote from other large objects.

Name	Mass ($\times 10^{20}$ kg)	Radius (km)
Vesta	2·59	263
Ceres	9·39	473

   (a) Dawn began orbiting Vesta, in a circular orbit, at a height of 680 km above the surface of the asteroid. The gravitational force acting on Dawn at this altitude was 24·1 N.

   (i) Show that the tangential velocity of Dawn in this orbit is 135 m s^{-1}. **2**

   Space for working and answer

   (ii) Calculate the orbital period of Dawn. **3**

   Space for working and answer

Page twelve

4. (continued)

   (b) Later in its mission, Dawn entered orbit around Ceres. It then moved from a high orbit to a lower orbit around the asteroid.

   (i) State what is meant by the *gravitational potential of a point in space*. **1**

   (ii) Dawn has a gravitational potential of $-1\cdot29 \times 10^4$ J kg^{-1} in the high orbit and a gravitational potential of $-3\cdot22 \times 10^4$ J kg^{-1} in the lower orbit.

   Determine the change in the potential energy of Dawn as a result of this change in orbit. **4**

   Space for working and answer

5. Two students are discussing objects escaping from the gravitational pull of the Earth. They make the following statements:

   Student 1: A rocket has to accelerate until it reaches the escape velocity of the Earth in order to escape its gravitational pull.

   Student 2: The moon is travelling slower than the escape velocity of the Earth and yet it has escaped.

   Use your knowledge of physics to comment on these statements. **3**

[Turn over for next question

DO NOT WRITE ON THIS PAGE

6. A Hertzsprung-Russell (H-R) diagram is shown in Figure 6A.

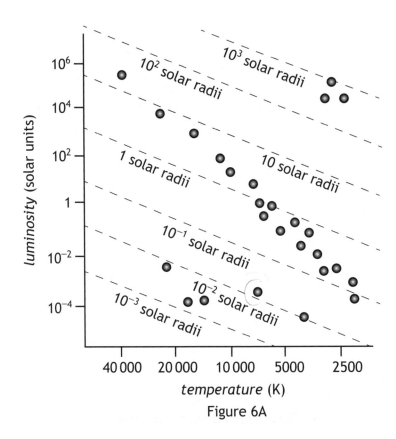

Figure 6A

(a) All stars on the main sequence release energy by converting hydrogen to helium. This process is known as the proton-proton (p-p) chain. One stage in the p-p chain is shown.

$$^{1}_{1}H + ^{1}_{1}H \rightarrow ^{2}_{1}H + x + y$$

Name particles x and y.

(b) The luminosity of the Sun is $3\cdot9 \times 10^{26}$ W. The star Procyon B has a luminosity of $4\cdot9 \times 10^{-4}$ solar units and a radius of $1\cdot2 \times 10^{-2}$ solar radii.

  (i) On the H-R diagram, circle the star at the position of Procyon B.

   (An additional diagram, if required, can be found on *Page forty-two*.)

  (ii) What type of star is Procyon B?

6. (b) (continued)

(iii) The apparent brightness of Procyon B when viewed from Earth is 1.3×10^{-12} W m^{-2}.

Calculate the distance of Procyon B from Earth. **4**

Space for working and answer

(c) The expression

$$\frac{L}{L_0} = 1.5 \left(\frac{M}{M_0}\right)^{3.5}$$

can be used to approximate the relationship between a star's mass M and its luminosity L.

L_0 is the luminosity of the Sun (1 solar unit) and M_0 is the mass of the Sun.

This expression is valid for stars of mass between $2M_0$ and $20M_0$.

Spica is a star which has mass $10.3M_0$.

Determine the approximate luminosity of Spica in solar units. **2**

Space for working and answer

7. Laser light is often described as having a single frequency. However, in practice a laser will emit photons with a range of frequencies.

Quantum physics links the frequency of a photon to its energy.

Therefore the photons emitted by a laser have a range of energies (ΔE). The range of photon energies is related to the lifetime (Δt) of the atom in the excited state.

A graph showing the variation of intensity with frequency for light from two types of laser is shown in Figure 7A.

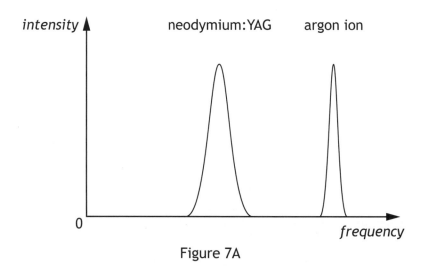

Figure 7A

(a) By considering the Heisenberg uncertainty principle, state how the lifetime of atoms in the excited state in the neodymium:YAG laser compares with the lifetime of atoms in the excited state in the argon ion laser.

Justify your answer. 2

7. (continued)

   (b) In another type of laser, an atom is in the excited state for a time of 5.0×10^{-6} s.

       (i) Calculate the minimum uncertainty in the energy (ΔE_{min}) of a photon emitted when the atom returns to its unexcited state.

       Space for working and answer

       (ii) Determine a value for the range of frequencies (Δf) of the photons emitted by this laser.

       Space for working and answer

8. A student is investigating simple harmonic motion. An oscillating mass on a spring, and a motion sensor connected to a computer, are used in the investigation. This is shown in Figure 8A.

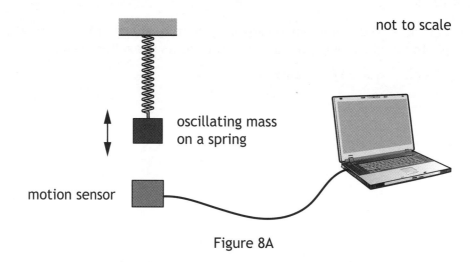

Figure 8A

The student raises the mass from its rest position and then releases it. The computer starts recording data when the mass is released.

(a) The student plans to model the displacement y of the mass from its rest position, using the expression

$$y = A \sin \omega t$$

where the symbols have their usual meaning.

Explain why the student is incorrect. **1**

8. (continued)

(b) (i) The unbalanced force acting on the mass is given by the expression

$$F = -m\omega^2 y$$

Hooke's Law is given by the expression

$$F = -ky$$

where k is the spring constant.

By comparing these expressions, show that the frequency of the oscillation can be described by the relationship

$$f = \frac{1}{2\pi}\sqrt{\frac{k}{m}}$$

2

(ii) The student measures the mass to be 0·50 kg and the period of oscillation to be 0·80 s.

Determine a value for the spring constant k.

Space for working and answer

3

8. (b) (continued)

(iii) The student plans to repeat the experiment using the same mass and a second spring, which has a spring constant twice the value of the original.

Determine the expected period of oscillation of the mass. **2**

(c) The student obtains graphs showing the variation of displacement with time, velocity with time and acceleration with time.

The student forgets to label the y-axis for each graph.

Complete the labelling of the y-axis of each graph in Figure 8B. **2**

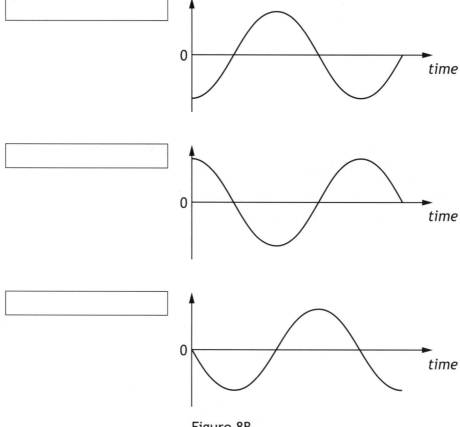

Figure 8B

9. A wave travelling along a string is represented by the relationship

$$y = 9 \cdot 50 \times 10^{-4} \sin(922t - 4 \cdot 50x)$$

(a) (i) Show that the frequency of the wave is 147 Hz.

Space for working and answer

(ii) Determine the speed of the wave.

Space for working and answer

(iii) The wave loses energy as it travels along the string.

At one point, the energy of the wave has decreased to one eighth of its original value.

Calculate the amplitude of the wave at this point.

Space for working and answer

9. (continued)

(b) The speed of a wave on a string can also be described by the relationship

$$v = \sqrt{\frac{T}{\mu}}$$

where v is the speed of the wave,

T is the tension in the string, and

μ is the mass per unit length of the string.

A string of length 0·69 m has a mass of $9·0 \times 10^{-3}$ kg.

A wave is travelling along the string with a speed of 203 m s^{-1}.

Calculate the tension in the string.

Space for working and answer

9. (continued)

(c) When a string is fixed at both ends and plucked, a stationary wave is produced.

    (i) Explain briefly, in terms of the superposition of waves, how the stationary wave is produced. **1**

    (ii) The string is vibrating at its fundamental frequency of 270 Hz and produces the stationary wave pattern shown in Figure 9A.

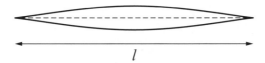

Figure 9A

Figure 9B shows the same string vibrating at a frequency called its third harmonic.

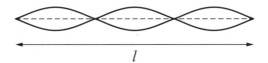

Figure 9B

Determine the frequency of the third harmonic. **1**

[BLANK PAGE]

10. The internal structure of some car windscreens produces an effect which can be likened to that obtained by slits in a grating.

   A passenger in a car observes a distant red traffic light and notices that the red light is surrounded by a pattern of bright spots.

   This is shown in Figure 10A.

   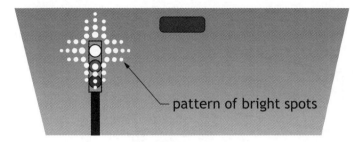

   Figure 10A

   (a) Explain how the **two-dimensional** pattern of bright spots shown in Figure 10A is produced. **2**

   (b) The traffic light changes to green. Apart from colour, state a difference that would be observed in the pattern of bright spots.

   Justify your answer. **2**

10. (continued)

(c) An LED from the traffic light is tested to determine the wavelength by shining its light through a set of Young's double slits, as shown in Figure 10B.

The fringe separation is $(13 \cdot 0 \pm 0 \cdot 5)$ mm and the double slit separation is $(0 \cdot 41 \pm 0 \cdot 01)$ mm.

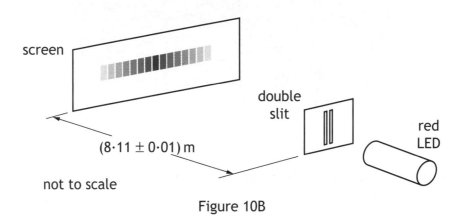

Figure 10B

(i) Calculate the wavelength of the light from the LED.   3

Space for working and answer

10. (c) (continued)

    (ii) Determine the absolute uncertainty in this wavelength. **5**

        Space for working and answer

    (iii) The experiment is now repeated with the screen moved further away from the slits.

        Explain why this is the most effective way of reducing the uncertainty in the calculated value of the wavelength. **1**

11. (a) State what is meant by the term *electric field strength*. **1**

(b) A, B, C and D are the vertices of a square of side 0·12 m.

Two +4·0 nC point charges are placed at positions B and D as shown in Figure 11A.

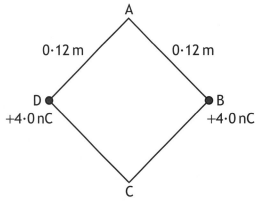

Figure 11A

(i) Show that the magnitude of the electric field strength at position A is $3·5 \times 10^3 \, NC^{-1}$. **3**

Space for working and answer

11. (b) (continued)

(ii) A +1·9 nC point charge is placed at position A.

Calculate the magnitude of the force acting on this charge. **3**

Space for working and answer

(iii) State the direction of the force acting on this charge. **1**

(iv) A fourth point charge is now placed at position C so that the resultant force on the charge at position A is zero.

Determine the magnitude of the charge placed at position C. **4**

Space for working and answer

12. A velocity selector is used as the initial part of a larger apparatus that is designed to measure properties of ions of different elements.

The velocity selector has a region in which there is a uniform electric field and a uniform magnetic field. These fields are perpendicular to each other and also perpendicular to the initial velocity v of the ions.

This is shown in Figure 12A.

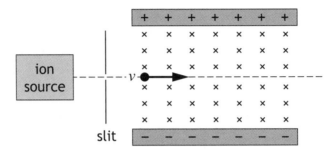

Figure 12A

A beam of chlorine ions consists of a number of isotopes including $^{35}Cl^+$.

The magnitude of the charge on a $^{35}Cl^+$ ion is $1 \cdot 60 \times 10^{-19}$ C.

The magnitude of electric force on a $^{35}Cl^+$ chlorine ion is $4 \cdot 00 \times 10^{-15}$ N.

The ions enter the apparatus with a range of speeds.

The magnetic induction is 115 mT.

(a) State the direction of the magnetic force on a $^{35}Cl^+$ ion. 1

(b) By considering the electric and magnetic forces acting on a $^{35}Cl^+$ ion, determine the speed of the $^{35}Cl^+$ ions that will pass through the apparatus without being deflected. 3

Space for working and answer

12. (continued)

(c) $^{35}Cl^+$ ions that are travelling at a velocity less than that determined in (b) are observed to follow the path shown in Figure 12B.

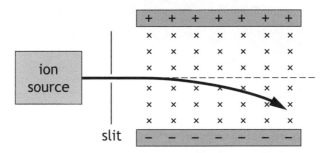

Figure 12B

Explain, in terms of their velocity, why these ions follow this path. **2**

(d) $^{37}Cl^{2+}$ ions are also present in the beam. $^{37}Cl^{2+}$ ions have a greater mass and a greater charge than $^{35}Cl^+$ ions. Some $^{37}Cl^{2+}$ ions also pass through the apparatus without being deflected.

State the speed of these ions.

You must justify your answer. **2**

13. A student purchases a capacitor with capacitance 1·0 F. The capacitor, which has negligible resistance, is used to supply short bursts of energy to the audio system in a car when there is high energy demand on the car battery.

The instructions state that the capacitor must be fully charged from the 12 V d.c. car battery through a 1·0 kΩ series resistor.

(a) Show that the time constant for this charging circuit is $1·0 \times 10^3$ s.

Space for working and answer

13. (continued)

    (b) The student carries out an experiment to monitor the voltage across the capacitor while it is being charged.

    (i) Draw a diagram of the circuit which would enable the student to carry out this experiment. **1**

    (ii) The student draws the graph shown in Figure 13A.

    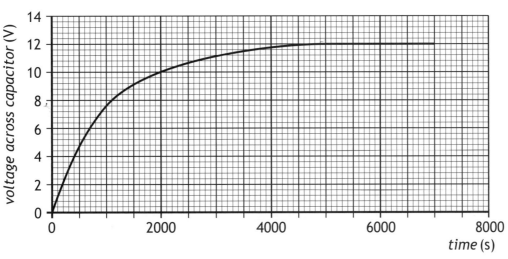

    Figure 13A

    (A) Use information from the graph to show that the capacitor is 63% charged after 1 time constant. **2**

    Space for working and answer

    (B) Use information from the graph to determine how many time constants are required for this capacitor to be considered fully charged. **1**

13. (continued)

(c) The car audio system is rated at 12 V, 20 W.

Use your knowledge of physics to comment on the suitability of the capacitor as the only energy source for the audio system.

3

[Turn over for next question

DO NOT WRITE ON THIS PAGE

14. A student designs a loudspeaker circuit.

   A capacitor and an inductor are used in the circuit so that high frequency signals are passed to a small "tweeter" loudspeaker and low frequency signals are passed to a large "woofer" loudspeaker.

   Each loudspeaker has a resistance of 8·0 Ω.

   The circuit diagram is shown in Figure 14A.

   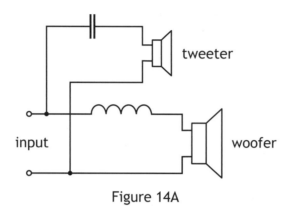

   Figure 14A

   The circuit is designed to have a "crossover" frequency of 3·0 kHz: at frequencies above 3·0 kHz there is a greater current in the tweeter and at frequencies below 3·0 kHz there is a greater current in the woofer.

   (a) Explain how the use of a capacitor and an inductor allows:

   (i) high frequency signals to be passed to the tweeter;

   (ii) low frequency signals to be passed to the woofer.

14. (continued)

(b) At the crossover frequency, both the reactance of the capacitor and the reactance of the inductor are equal to the resistance of each loudspeaker.

Calculate the inductance required to provide an inductive reactance of $8 \cdot 0\,\Omega$ when the frequency of the signal is $3 \cdot 0\,\text{kHz}$.

Space for working and answer

3

14. (continued)

(c) In a box of components, the student finds an inductor and decides to determine its inductance. The student constructs the circuit shown in Figure 14B.

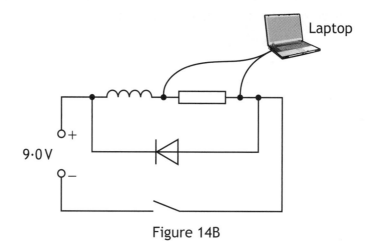

Figure 14B

The student obtains data from the experiment and presents the data on the graph shown in Figure 14C.

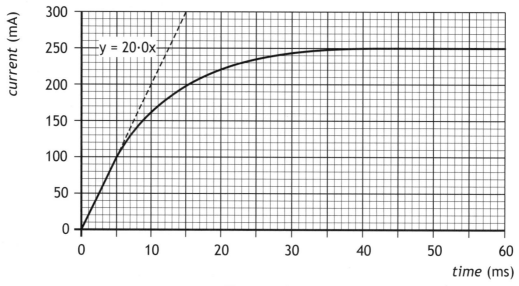

Figure 14C

14. (c) (continued)

    (i) Determine the inductance of the inductor. 4

    Space for working and answer

    (ii) The student was advised to include a diode in the circuit to prevent damage to the laptop when the switch is opened.

    Explain why this is necessary. 1

[END OF QUESTION PAPER]

ADDITIONAL SPACE FOR ANSWERS AND ROUGH WORK

Additional diagram for question 2 (c)

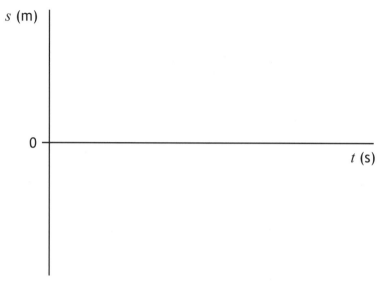

Figure 2G

Additional diagram for question 6 (b) (i)

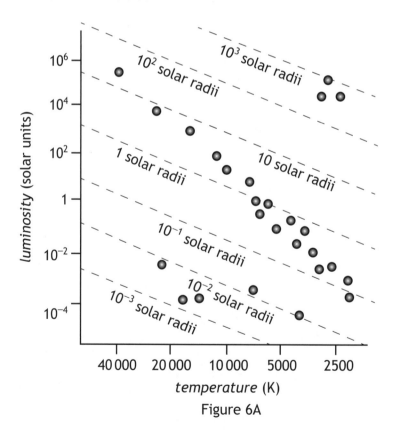

Figure 6A

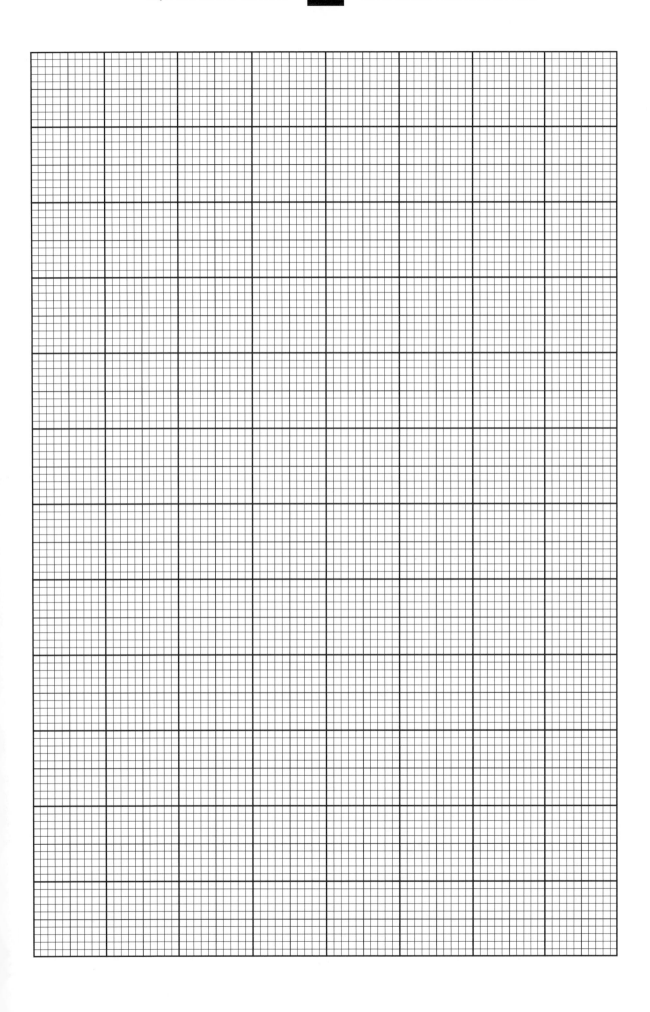

ADDITIONAL SPACE FOR ANSWERS AND ROUGH WORK

ADVANCED HIGHER
2018

X757/77/11

National Qualifications 2018

**Physics
Relationships Sheet**

TUESDAY, 8 MAY
9:00 AM – 11:30 AM

Relationships required for Physics Advanced Higher

$v = \dfrac{ds}{dt}$

$a = \dfrac{dv}{dt} = \dfrac{d^2s}{dt^2}$

$v = u + at$

$s = ut + \dfrac{1}{2}at^2$

$v^2 = u^2 + 2as$

$\omega = \dfrac{d\theta}{dt}$

$\alpha = \dfrac{d\omega}{dt} = \dfrac{d^2\theta}{dt^2}$

$\omega = \omega_o + \alpha t$

$\theta = \omega_o t + \dfrac{1}{2}\alpha t^2$

$\omega^2 = \omega_o^2 + 2\alpha\theta$

$s = r\theta$

$v = r\omega$

$a_t = r\alpha$

$a_r = \dfrac{v^2}{r} = r\omega^2$

$F = \dfrac{mv^2}{r} = mr\omega^2$

$T = Fr$

$T = I\alpha$

$L = mvr = mr^2\omega$

$L = I\omega$

$E_K = \dfrac{1}{2}I\omega^2$

$F = G\dfrac{Mm}{r^2}$

$V = -\dfrac{GM}{r}$

$v = \sqrt{\dfrac{2GM}{r}}$

apparent brightness, $b = \dfrac{L}{4\pi r^2}$

Power per unit area $= \sigma T^4$

$L = 4\pi r^2 \sigma T^4$

$r_{Schwarzschild} = \dfrac{2GM}{c^2}$

$E = hf$

$\lambda = \dfrac{h}{p}$

$mvr = \dfrac{nh}{2\pi}$

$\Delta x \, \Delta p_x \geq \dfrac{h}{4\pi}$

$\Delta E \, \Delta t \geq \dfrac{h}{4\pi}$

$F = qvB$

$\omega = 2\pi f$

$\omega = \dfrac{2\pi}{T}$

$$a = \frac{d^2y}{dt^2} = -\omega^2 y$$

$$y = A\cos\omega t \quad \text{or} \quad y = A\sin\omega t$$

$$v = \pm\omega\sqrt{(A^2 - y^2)}$$

$$E_K = \frac{1}{2}m\omega^2(A^2 - y^2)$$

$$E_P = \frac{1}{2}m\omega^2 y^2$$

$$y = A\sin 2\pi(ft - \frac{x}{\lambda})$$

$$E = kA^2$$

$$\phi = \frac{2\pi x}{\lambda}$$

optical path difference $= m\lambda$ or $\left(m+\frac{1}{2}\right)\lambda$

where $m = 0, 1, 2....$

$$\Delta x = \frac{\lambda l}{2d}$$

$$d = \frac{\lambda}{4n}$$

$$\Delta x = \frac{\lambda D}{d}$$

$$n = \tan i_P$$

$$F = \frac{Q_1 Q_2}{4\pi\varepsilon_o r^2}$$

$$E = \frac{Q}{4\pi\varepsilon_o r^2}$$

$$V = \frac{Q}{4\pi\varepsilon_o r}$$

$$F = QE$$

$$V = Ed$$

$$F = IlB\sin\theta$$

$$B = \frac{\mu_o I}{2\pi r}$$

$$c = \frac{1}{\sqrt{\varepsilon_o \mu_o}}$$

$$t = RC$$

$$X_C = \frac{V}{I}$$

$$X_C = \frac{1}{2\pi fC}$$

$$\varepsilon = -L\frac{dI}{dt}$$

$$E = \frac{1}{2}LI^2$$

$$X_L = \frac{V}{I}$$

$$X_L = 2\pi fL$$

$$\frac{\Delta W}{W} = \sqrt{\left(\frac{\Delta X}{X}\right)^2 + \left(\frac{\Delta Y}{Y}\right)^2 + \left(\frac{\Delta Z}{Z}\right)^2}$$

$$\Delta W = \sqrt{\Delta X^2 + \Delta Y^2 + \Delta Z^2}$$

$d = \bar{v}t$

$s = \bar{v}t$

$v = u + at$

$s = ut + \frac{1}{2}at^2$

$v^2 = u^2 + 2as$

$s = \frac{1}{2}(u+v)t$

$W = mg$

$F = ma$

$E_W = Fd$

$E_P = mgh$

$E_K = \frac{1}{2}mv^2$

$P = \frac{E}{t}$

$p = mv$

$Ft = mv - mu$

$F = G\frac{Mm}{r^2}$

$t' = \dfrac{t}{\sqrt{1-\left(v/c\right)^2}}$

$l' = l\sqrt{1-\left(v/c\right)^2}$

$f_o = f_s \left(\dfrac{v}{v \pm v_s}\right)$

$z = \dfrac{\lambda_{observed} - \lambda_{rest}}{\lambda_{rest}}$

$z = \dfrac{v}{c}$

$v = H_0 d$

$W = QV$

$E = mc^2$

$E = hf$

$E_K = hf - hf_0$

$E_2 - E_1 = hf$

$T = \dfrac{1}{f}$

$v = f\lambda$

$d\sin\theta = m\lambda$

$n = \dfrac{\sin\theta_1}{\sin\theta_2}$

$\dfrac{\sin\theta_1}{\sin\theta_2} = \dfrac{\lambda_1}{\lambda_2} = \dfrac{v_1}{v_2}$

$\sin\theta_c = \dfrac{1}{n}$

$I = \dfrac{k}{d^2}$

$I = \dfrac{P}{A}$

path difference $= m\lambda$ or $\left(m + \dfrac{1}{2}\right)\lambda$ where $m = 0, 1, 2\ldots$

random uncertainty $= \dfrac{\text{max. value} - \text{min. value}}{\text{number of values}}$

$V_{peak} = \sqrt{2}V_{rms}$

$I_{peak} = \sqrt{2}I_{rms}$

$Q = It$

$V = IR$

$P = IV = I^2R = \dfrac{V^2}{R}$

$R_T = R_1 + R_2 + \ldots$

$\dfrac{1}{R_T} = \dfrac{1}{R_1} + \dfrac{1}{R_2} + \ldots$

$E = V + Ir$

$V_1 = \left(\dfrac{R_1}{R_1 + R_2}\right)V_S$

$\dfrac{V_1}{V_2} = \dfrac{R_1}{R_2}$

$C = \dfrac{Q}{V}$

$E = \dfrac{1}{2}QV = \dfrac{1}{2}CV^2 = \dfrac{1}{2}\dfrac{Q^2}{C}$

Additional Relationships

Circle

circumference = $2\pi r$

area = πr^2

Sphere

area = $4\pi r^2$

volume = $\frac{4}{3}\pi r^3$

Trigonometry

$\sin \theta = \dfrac{\text{opposite}}{\text{hypotenuse}}$

$\cos \theta = \dfrac{\text{adjacent}}{\text{hypotenuse}}$

$\tan \theta = \dfrac{\text{opposite}}{\text{adjacent}}$

$\sin^2 \theta + \cos^2 \theta = 1$

Moment of inertia

point mass
$I = mr^2$

rod about centre
$I = \frac{1}{12} ml^2$

rod about end
$I = \frac{1}{3} ml^2$

disc about centre
$I = \frac{1}{2} mr^2$

sphere about centre
$I = \frac{2}{5} mr^2$

Table of standard derivatives

$f(x)$	$f'(x)$
$\sin ax$	$a \cos ax$
$\cos ax$	$-a \sin ax$

Table of standard integrals

$f(x)$	$\int f(x)dx$
$\sin ax$	$-\dfrac{1}{a} \cos ax + C$
$\cos ax$	$\dfrac{1}{a} \sin ax + C$

Electron Arrangements of Elements

Key

| Atomic number |
| Symbol |
| Electron arrangement |
| Name |

Main Groups

Group 1 (1)	Group 2 (2)		Group 3 (13)	Group 4 (14)	Group 5 (15)	Group 6 (16)	Group 7 (17)	Group 0 (18)
1 **H** 1 Hydrogen								2 **He** 2 Helium
3 **Li** 2,1 Lithium	4 **Be** 2,2 Beryllium		5 **B** 2,3 Boron	6 **C** 2,4 Carbon	7 **N** 2,5 Nitrogen	8 **O** 2,6 Oxygen	9 **F** 2,7 Fluorine	10 **Ne** 2,8 Neon
11 **Na** 2,8,1 Sodium	12 **Mg** 2,8,2 Magnesium		13 **Al** 2,8,3 Aluminium	14 **Si** 2,8,4 Silicon	15 **P** 2,8,5 Phosphorus	16 **S** 2,8,6 Sulfur	17 **Cl** 2,8,7 Chlorine	18 **Ar** 2,8,8 Argon

Transition Elements

	(3)	(4)	(5)	(6)	(7)	(8)	(9)	(10)	(11)	(12)						
19 **K** 2,8,8,1 Potassium	20 **Ca** 2,8,8,2 Calcium	21 **Sc** 2,8,9,2 Scandium	22 **Ti** 2,8,10,2 Titanium	23 **V** 2,8,11,2 Vanadium	24 **Cr** 2,8,13,1 Chromium	25 **Mn** 2,8,13,2 Manganese	26 **Fe** 2,8,14,2 Iron	27 **Co** 2,8,15,2 Cobalt	28 **Ni** 2,8,16,2 Nickel	29 **Cu** 2,8,18,1 Copper	30 **Zn** 2,8,18,2 Zinc	31 **Ga** 2,8,18,3 Gallium	32 **Ge** 2,8,18,4 Germanium	33 **As** 2,8,18,5 Arsenic	34 **Se** 2,8,18,6 Selenium	35 **Br** 2,8,18,7 Bromine	36 **Kr** 2,8,18,8 Krypton
37 **Rb** 2,8,18,8,1 Rubidium	38 **Sr** 2,8,18,8,2 Strontium	39 **Y** 2,8,18,9,2 Yttrium	40 **Zr** 2,8,18,10,2 Zirconium	41 **Nb** 2,8,18,12,1 Niobium	42 **Mo** 2,8,18,13,1 Molybdenum	43 **Tc** 2,8,18,13,2 Technetium	44 **Ru** 2,8,18,15,1 Ruthenium	45 **Rh** 2,8,18,16,1 Rhodium	46 **Pd** 2,8,18,18,0 Palladium	47 **Ag** 2,8,18,18,1 Silver	48 **Cd** 2,8,18,18,2 Cadmium	49 **In** 2,8,18,18,3 Indium	50 **Sn** 2,8,18,18,4 Tin	51 **Sb** 2,8,18,18,5 Antimony	52 **Te** 2,8,18,18,6 Tellurium	53 **I** 2,8,18,18,7 Iodine	54 **Xe** 2,8,18,18,8 Xenon
55 **Cs** 2,8,18,18,8,1 Caesium	56 **Ba** 2,8,18,18,8,2 Barium	57 **La** 2,8,18,18,9,2 Lanthanum	72 **Hf** 2,8,18,32,10,2 Hafnium	73 **Ta** 2,8,18,32,11,2 Tantalum	74 **W** 2,8,18,32,12,2 Tungsten	75 **Re** 2,8,18,32,13,2 Rhenium	76 **Os** 2,8,18,32,14,2 Osmium	77 **Ir** 2,8,18,32,15,2 Iridium	78 **Pt** 2,8,18,32,17,1 Platinum	79 **Au** 2,8,18,32,18,1 Gold	80 **Hg** 2,8,18,32,18,2 Mercury	81 **Tl** 2,8,18,32,18,3 Thallium	82 **Pb** 2,8,18,32,18,4 Lead	83 **Bi** 2,8,18,32,18,5 Bismuth	84 **Po** 2,8,18,32,18,6 Polonium	85 **At** 2,8,18,32,18,7 Astatine	86 **Rn** 2,8,18,32,18,8 Radon
87 **Fr** 2,8,18,32,18,8,1 Francium	88 **Ra** 2,8,18,32,18,8,2 Radium	89 **Ac** 2,8,18,32,18,9,2 Actinium	104 **Rf** 2,8,18,32,32,10,2 Rutherfordium	105 **Db** 2,8,18,32,32,11,2 Dubnium	106 **Sg** 2,8,18,32,32,12,2 Seaborgium	107 **Bh** 2,8,18,32,32,13,2 Bohrium	108 **Hs** 2,8,18,32,32,14,2 Hassium	109 **Mt** 2,8,18,32,32,15,2 Meitnerium	110 **Ds** 2,8,18,32,32,17,1 Darmstadtium	111 **Rg** 2,8,18,32,32,18,1 Roentgenium	112 **Cn** 2,8,18,32,32,18,2 Copernicium						

Lanthanides

57 **La** 2,8,18,18,9,2 Lanthanum	58 **Ce** 2,8,18,20,8,2 Cerium	59 **Pr** 2,8,18,21,8,2 Praseodymium	60 **Nd** 2,8,18,22,8,2 Neodymium	61 **Pm** 2,8,18,23,8,2 Promethium	62 **Sm** 2,8,18,24,8,2 Samarium	63 **Eu** 2,8,18,25,8,2 Europium	64 **Gd** 2,8,18,25,9,2 Gadolinium	65 **Tb** 2,8,18,27,8,2 Terbium	66 **Dy** 2,8,18,28,8,2 Dysprosium	67 **Ho** 2,8,18,29,8,2 Holmium	68 **Er** 2,8,18,30,8,2 Erbium	69 **Tm** 2,8,18,31,8,2 Thulium	70 **Yb** 2,8,18,32,8,2 Ytterbium	71 **Lu** 2,8,18,32,9,2 Lutetium

Actinides

89 **Ac** 2,8,18,32,18,9,2 Actinium	90 **Th** 2,8,18,32,18,10,2 Thorium	91 **Pa** 2,8,18,32,20,9,2 Protactinium	92 **U** 2,8,18,32,21,9,2 Uranium	93 **Np** 2,8,18,32,22,9,2 Neptunium	94 **Pu** 2,8,18,32,24,8,2 Plutonium	95 **Am** 2,8,18,32,25,8,2 Americium	96 **Cm** 2,8,18,32,25,9,2 Curium	97 **Bk** 2,8,18,32,27,8,2 Berkelium	98 **Cf** 2,8,18,32,28,8,2 Californium	99 **Es** 2,8,18,32,29,8,2 Einsteinium	100 **Fm** 2,8,18,32,30,8,2 Fermium	101 **Md** 2,8,18,32,31,8,2 Mendelevium	102 **No** 2,8,18,32,32,8,2 Nobelium	103 **Lr** 2,8,18,32,32,9,2 Lawrencium

AH

National Qualifications 2018

FOR OFFICIAL USE

Mark

X757/77/01

Physics

TUESDAY, 8 MAY
9:00 AM – 11:30 AM

Fill in these boxes and read what is printed below.

Full name of centre

Town

Forename(s)

Surname

Number of seat

Date of birth
Day Month Year

Scottish candidate number

Total marks — 140

Attempt ALL questions.

Reference may be made to the Physics Relationships Sheet X757/77/11 and the Data Sheet on *Page two*.

Write your answers clearly in the spaces provided in this booklet. Additional space for answers and rough work is provided at the end of this booklet. If you use this space you must clearly identify the question number you are attempting. Any rough work must be written in this booklet. You should score through your rough work when you have written your final copy.

Care should be taken to give an appropriate number of significant figures in the final answers to calculations.

Use **blue** or **black** ink.

Before leaving the examination room you must give this booklet to the Invigilator; if you do not, you may lose all the marks for this paper.

SQA

DATA SHEET

COMMON PHYSICAL QUANTITIES

Quantity	Symbol	Value	Quantity	Symbol	Value
Gravitational acceleration on Earth	g	$9{\cdot}8$ m s^{-2}	Mass of electron	m_e	$9{\cdot}11 \times 10^{-31}$ kg
Radius of Earth	R_E	$6{\cdot}4 \times 10^6$ m	Charge on electron	e	$-1{\cdot}60 \times 10^{-19}$ C
Mass of Earth	M_E	$6{\cdot}0 \times 10^{24}$ kg	Mass of neutron	m_n	$1{\cdot}675 \times 10^{-27}$ kg
Mass of Moon	M_M	$7{\cdot}3 \times 10^{22}$ kg	Mass of proton	m_p	$1{\cdot}673 \times 10^{-27}$ kg
Radius of Moon	R_M	$1{\cdot}7 \times 10^6$ m	Mass of alpha particle	m_α	$6{\cdot}645 \times 10^{-27}$ kg
Mean Radius of Moon Orbit		$3{\cdot}84 \times 10^8$ m	Charge on alpha particle		$3{\cdot}20 \times 10^{-19}$ C
Solar radius		$6{\cdot}955 \times 10^8$ m	Planck's constant	h	$6{\cdot}63 \times 10^{-34}$ J s
Mass of Sun		$2{\cdot}0 \times 10^{30}$ kg	Permittivity of free space	ε_0	$8{\cdot}85 \times 10^{-12}$ F m^{-1}
1 AU		$1{\cdot}5 \times 10^{11}$ m	Permeability of free space	μ_0	$4\pi \times 10^{-7}$ H m^{-1}
Stefan-Boltzmann constant	σ	$5{\cdot}67 \times 10^{-8}$ W m^{-2} K^{-4}	Speed of light in vacuum	c	$3{\cdot}00 \times 10^8$ m s^{-1}
Universal constant of gravitation	G	$6{\cdot}67 \times 10^{-11}$ m^3 kg^{-1} s^{-2}	Speed of sound in air	v	$3{\cdot}4 \times 10^2$ m s^{-1}

REFRACTIVE INDICES

The refractive indices refer to sodium light of wavelength 589 nm and to substances at a temperature of 273 K.

Substance	Refractive index	Substance	Refractive index
Diamond	2·42	Glycerol	1·47
Glass	1·51	Water	1·33
Ice	1·31	Air	1·00
Perspex	1·49	Magnesium Fluoride	1·38

SPECTRAL LINES

Element	Wavelength/nm	Colour	Element	Wavelength/nm	Colour
Hydrogen	656	Red	Cadmium	644	Red
	486	Blue-green		509	Green
	434	Blue-violet		480	Blue
	410	Violet	Lasers		
	397	Ultraviolet	Element	Wavelength/nm	Colour
	389	Ultraviolet	Carbon dioxide	9550 } 10590 }	Infrared
Sodium	589	Yellow	Helium-neon	633	Red

PROPERTIES OF SELECTED MATERIALS

Substance	Density/ kg m^{-3}	Melting Point/ K	Boiling Point/ K	Specific Heat Capacity/ J kg^{-1} K^{-1}	Specific Latent Heat of Fusion/ J kg^{-1}	Specific Latent Heat of Vaporisation/ J kg^{-1}
Aluminium	$2{\cdot}70 \times 10^3$	933	2623	$9{\cdot}02 \times 10^2$	$3{\cdot}95 \times 10^5$	....
Copper	$8{\cdot}96 \times 10^3$	1357	2853	$3{\cdot}86 \times 10^2$	$2{\cdot}05 \times 10^5$	....
Glass	$2{\cdot}60 \times 10^3$	1400	....	$6{\cdot}70 \times 10^2$	....	....
Ice	$9{\cdot}20 \times 10^2$	273	....	$2{\cdot}10 \times 10^3$	$3{\cdot}34 \times 10^5$	....
Glycerol	$1{\cdot}26 \times 10^3$	291	563	$2{\cdot}43 \times 10^3$	$1{\cdot}81 \times 10^5$	$8{\cdot}30 \times 10^5$
Methanol	$7{\cdot}91 \times 10^2$	175	338	$2{\cdot}52 \times 10^3$	$9{\cdot}9 \times 10^4$	$1{\cdot}12 \times 10^6$
Sea Water	$1{\cdot}02 \times 10^3$	264	377	$3{\cdot}93 \times 10^3$	....	....
Water	$1{\cdot}00 \times 10^3$	273	373	$4{\cdot}18 \times 10^3$	$3{\cdot}34 \times 10^5$	$2{\cdot}26 \times 10^6$
Air	1·29	....	....	....	....	....
Hydrogen	$9{\cdot}0 \times 10^{-2}$	14	20	$1{\cdot}43 \times 10^4$	....	$4{\cdot}50 \times 10^5$
Nitrogen	1·25	63	77	$1{\cdot}04 \times 10^3$	....	$2{\cdot}00 \times 10^5$
Oxygen	1·43	55	90	$9{\cdot}18 \times 10^2$	....	$2{\cdot}40 \times 10^4$

The gas densities refer to a temperature of 273 K and a pressure of $1{\cdot}01 \times 10^5$ Pa.

Total marks — 140

Attempt ALL questions

1. Energy is stored in a clockwork toy car by winding-up an internal spring using a key. The car is shown in Figure 1A.

   Figure 1A

   The car is released on a horizontal surface and moves forward in a straight line. It eventually comes to rest.

   The velocity v of the car, at time t after its release, is given by the relationship

   $$v = 0 \cdot 0071t - 0 \cdot 00025t^2$$

   where v is measured in m s^{-1} and t is measured in s.

   Using calculus methods:

   (a) determine the acceleration of the car 20·0 s after its release;   3

   Space for working and answer

   (b) determine the distance travelled by the car 20·0 s after its release.   3

   Space for working and answer

2. (a) A student places a radio-controlled car on a horizontal circular track, as shown in Figure 2A.

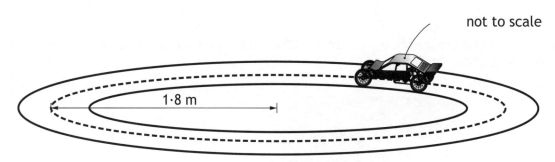

Figure 2A

The car travels around the track with a constant speed of $3.5\,\mathrm{m\,s^{-1}}$. The track has a radius of $1.8\,\mathrm{m}$.

(i) Explain why the car is accelerating, even though it is travelling at a constant speed. 1

(ii) Calculate the radial acceleration of the car. 3

Space for working and answer

2. (a) (continued)

   (iii) The car has a mass of 0·431 kg.

   The student now increases the speed of the car to 5·5 m s^{-1}.

   The total radial friction between the car and the track has a maximum value of 6·4 N.

   Show by calculation that the car cannot continue to travel in a circular path. **3**

   Space for working and answer

[Turn over

2. (continued)

(b) The car is now placed on a track, which includes a raised section. This is shown in Figure 2B.

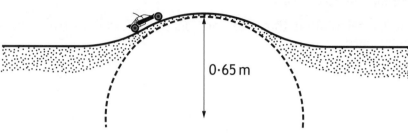

Figure 2B

The raised section of the track can be considered as the arc of a circle, which has radius r of 0·65 m.

(i) The car will lose contact with the raised section of track if its speed is greater than v_{max}.

Show that v_{max} is given by the relationship

$$v_{max} = \sqrt{gr}$$

(ii) Calculate the maximum speed v_{max} at which the car can cross the raised section without losing contact with the track.

Space for working and answer

2. (b) (continued)

   (iii) A second car, with a smaller mass than the first car, approaches the raised section at the same speed as calculated in (b)(ii).

   State whether the second car will lose contact with the track as it crosses the raised section.

   Justify your answer in terms of forces acting on the car. 2

[Turn over

3. Wheels on road vehicles can vibrate if the wheel is not 'balanced'. Garages can check that each wheel is balanced using a wheel balancing machine, as shown in Figure 3A.

Figure 3A

The wheel is rotated about its axis by the wheel balancing machine.

The angular velocity of the wheel increases uniformly from rest with an angular acceleration of 6.7 rad s^{-2}.

(a) The wheel reaches its maximum angular velocity after 3.9 s.

Show that its maximum angular velocity is 26 rad s^{-1}.

Space for working and answer

3. (continued)

(b) After 3·9 s, the rotational kinetic energy of the wheel is 430 J.

Calculate the moment of inertia of the wheel. 3

Space for working and answer

(c) A brake is applied which brings the wheel uniformly to rest from its maximum velocity.

The wheel completes 14 revolutions during the braking process.

(i) Calculate the angular acceleration of the wheel during the braking process. 4

Space for working and answer

(ii) Calculate the braking torque applied by the wheel balancing machine. 3

Space for working and answer

4. Astronomers have discovered another solar system in our galaxy. The main sequence star, HD 69830, lies at the centre of this solar system. This solar system also includes three exoplanets, b, c, and d and an asteroid belt.

This solar system is shown in Figure 4A.

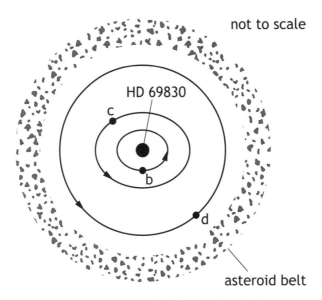

Figure 4A

(a) The orbit of exoplanet d can be considered circular.

To a reasonable approximation the centripetal force on exoplanet d is provided by the gravitational attraction of star HD 69830.

(i) Show that, for a circular orbit of radius r, the period T of a planet about a parent star of mass M, is given by

$$T^2 = \frac{4\pi^2}{GM} r^3$$

3

4. (a) (continued)

(ii) Some information about this solar system is shown in the table below.

Exoplanet	Type of orbit	Mass in Earth masses	Mean orbital radius in Astronomical Units (AU)	Orbital period In Earth days
b	Elliptical	10·2	-	8·67
c	Elliptical	11·8	0·186	-
d	Circular	18·1	0·63	197

Determine the mass, in kg, of star HD 69830.

Space for working and answer

(b) Two asteroids collide at a distance of $1·58 \times 10^{11}$ m from the centre of the star HD 69830. As a result of this collision, one of the asteroids escapes from this solar system.

Calculate the minimum speed which this asteroid must have immediately after the collision, in order to escape from this solar system.

Space for working and answer

5. (a) Explain what is meant by the term *Schwarzschild radius*. 1

(b) (i) Calculate the Schwarzschild radius of the Sun. 3

Space for working and answer

(ii) Explain, with reference to its radius, why the Sun is not a black hole. 1

(c) The point of closest approach of a planet to the Sun is called the perihelion of the planet. The perihelion of Mercury rotates slowly around the Sun, as shown in Figure 5A.

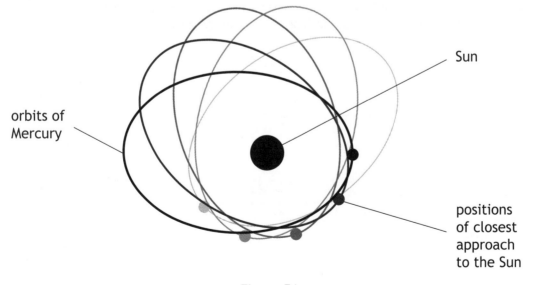

Figure 5A

5. (c) (continued)

This rotation of the perihelion is referred to as the precession of Mercury, and is due to the curvature of spacetime. This causes an angular change in the perihelion of Mercury.

The angular change **per orbit** is calculated using the relationship

$$\phi = 3\pi \frac{r_s}{a(1-e^2)}$$

where:

ϕ is the angular change **per orbit**, in radians;

r_s is the Schwarzschild radius of the Sun, in metres;

a is the semi-major axis of the orbit, for Mercury $a = 5 \cdot 805 \times 10^{10}$ m;

e is the eccentricity of the orbit, for Mercury $e = 0 \cdot 206$.

Mercury completes **four** orbits of the Sun in one Earth year.

Determine the angular change in the perihelion of Mercury **after one Earth year**.

Space for working and answer

6. Bellatrix and Acrab are two stars which are similar in size. However, the apparent brightness of each is different.

Use your knowledge of stellar physics to comment on why there is a difference in the apparent brightness of the two stars. **3**

[Turn over for next question

DO NOT WRITE ON THIS PAGE

7. In a crystal lattice, atoms are arranged in planes with a small gap between each plane.

   Neutron diffraction is a process which allows investigation of the structure of crystal lattices.

   In this process there are three stages:
   neutrons are accelerated;
   the neutrons pass through the crystal lattice;
   an interference pattern is produced.

   (a) (i) In this process, neutrons exhibit wave-particle duality.

   Identify the stage of the process which provides evidence for particle-like behaviour of neutrons. **1**

   (ii) Neutrons, each with a measured momentum of 1.29×10^{-23} kg m s^{-1} produce an observable interference pattern from one type of crystal lattice.

   Calculate the wavelength of a neutron travelling with this momentum. **3**

   Space for working and answer

   (iii) Explain the implication of the Heisenberg uncertainty principle for the precision of these experimental measurements. **1**

7. **(a)** **(continued)**

(iv) The momentum of a neutron is measured to be $1\cdot29 \times 10^{-23}$ kg m s^{-1} with a precision of $\pm\,3\cdot0\%$.

Determine the minimum **absolute** uncertainty in the position Δx_{min} of this neutron.

Space for working and answer

(b) Some of the neutrons used to investigate the structure of crystal lattices will not produce an observed interference pattern. This may be due to a large uncertainty in their momentum.

Explain why a large uncertainty in their momentum would result in these neutrons being unsuitable for this diffraction process.

8. (a) Inside the core of stars like the Sun, hydrogen nuclei fuse together to form heavier nuclei.

   (i) State the region of the Hertzsprung-Russell diagram in which stars like the Sun are located. **1**

   (ii) One type of fusion reaction is known as the proton-proton chain and is described below.

   $$6\,^1_1H \rightarrow \,^4_2X + 2\,^0_1Z + 2\,^0_0\nu + 2\,^1_1H + 2\,^0_0\gamma$$

   Identify the particles indicated by the letters X and Z. **2**

(b) High energy charged particles are ejected from the Sun.

   State the name given to the constant stream of charged particles which the Sun ejects. **1**

8. (continued)

(c) The stream of particles being ejected from the Sun produces an outward pressure. This outward pressure depends on the number of particles being ejected from the Sun and the speed of these particles.

The pressure at a distance of one astronomical unit (AU) from the Sun is given by the relationship

$$p = 1\cdot 6726 \times 10^{-6} \times n \times v^2$$

where:

p is the pressure in nanopascals;

n is the number of particles per cubic centimetre;

v is the speed of particles in kilometres per second.

(i) On one occasion, a pressure of $9\cdot 56 \times 10^{-10}$ Pa was recorded when the particle speed was measured to be $6\cdot 02 \times 10^5$ m s^{-1}.

Calculate the number of particles per cubic centimetre. **2**

Space for working and answer

(ii) The pressure decreases as the particles stream further from the Sun.

This is because the number of particles per cubic centimetre decreases and the kinetic energy of the particles decreases.

(A) Explain why the number of particles per cubic centimetre decreases as the particles stream further from the Sun. **1**

(B) Explain why the kinetic energy of the particles decreases as the particles stream further from the Sun. **1**

8. (continued)

(d) When the charged particles approach the Earth, the magnetic field of the Earth causes them to follow a helical path, as shown in Figure 8A.

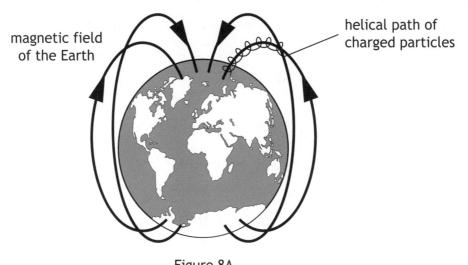

Figure 8A

Explain why the charged particles follow a helical path. 2

[Turn over for next question

DO NOT WRITE ON THIS PAGE

9. A ball-bearing is released from height h on a smooth curved track, as shown in Figure 9A.

The ball-bearing oscillates on the track about position P.

The motion of the ball-bearing can be modelled as Simple Harmonic Motion (SHM).

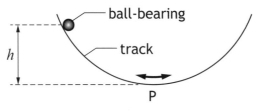

Figure 9A

(a) The ball-bearing makes 1·5 oscillations in 2·5 s.

(i) Show that the angular frequency of the ball-bearing is 3·8 rad s^{-1}.   **2**

Space for working and answer

(ii) The horizontal displacement x of the ball-bearing from position P at time t can be predicted using the relationship

$$x = -0\cdot2\cos(3\cdot8t)$$

Using calculus methods, show that this relationship is consistent with SHM.   **3**

9. (a) (continued)

(iii) Determine the maximum speed of the ball-bearing during its motion. **3**

Space for working and answer

(iv) Determine the height h from which the ball bearing was released. **3**

Space for working and answer

9. (continued)

(b) In practice, the maximum horizontal displacement of the ball-bearing decreases with time.

A graph showing the variation in the horizontal displacement of the ball-bearing from position P with time is shown in Figure 9B.

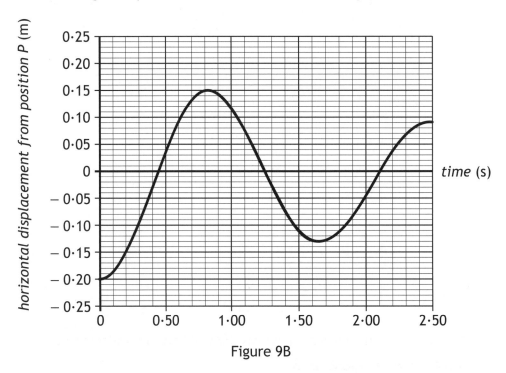

Figure 9B

Sketch a graph showing how the **vertical** displacement of the ball-bearing from position P changes over the same time period.

Numerical values are not required on either axis.

2

[Turn over for next question

DO NOT WRITE ON THIS PAGE

10. An electromagnetic wave is travelling along an optical fibre. Inside the fibre the electric field vectors oscillate, as shown in Figure 10A.

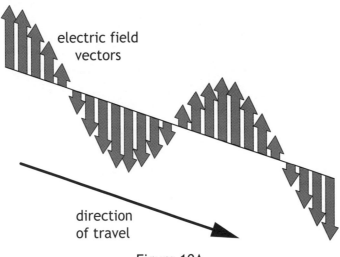

Figure 10A

The direction of travel of the wave is taken to be the x-direction.

The magnitude of the electric field vector E at any point x and time t is given by the relationship

$$E = 12 \times 10^{-6} \sin 2\pi \left(1 \cdot 31 \times 10^{14} t - \frac{x}{1 \cdot 55 \times 10^{-6}} \right)$$

(a) (i) Two points, A and B, along the wave are separated by a distance of $4 \cdot 25 \times 10^{-7}$ m in the x-direction.

Calculate the phase difference between points A and B.

Space for working and answer

10. (a) (continued)

(ii) Another two points on the wave, P and Q, have a phase difference of π radians.

State how the direction of the electric field vector at point P compares to the direction of the electric field vector at point Q. **1**

(b) (i) Show that the speed of the electromagnetic wave in this optical fibre is $2 \cdot 03 \times 10^8 \, \text{m s}^{-1}$. **2**

Space for working and answer

(ii) The speed v_m of an electromagnetic wave in a medium is given by the relationship

$$v_m = \frac{1}{\sqrt{\varepsilon_m \mu_m}}$$

The permeability μ_m of the optical fibre material can be considered to be equal to the permeability of free space.

By considering the speed of the electromagnetic wave in this fibre, determine the permittivity ε_m of the optical fibre material. **2**

Space for working and answer

11. A thin air wedge is formed between two glass plates of length 75 mm, which are in contact at one end and separated by a thin metal wire at the other end.

Figure 11A shows sodium light being reflected down onto the air wedge.

A travelling microscope is used to view the resulting interference pattern.

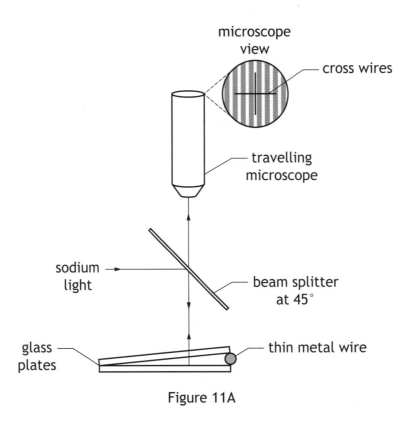

Figure 11A

A student observes the image shown in Figure 11B.

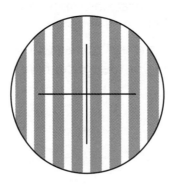

Figure 11B

The student aligns the cross-hairs to a bright fringe and then moves the travelling microscope until 20 further bright fringes have passed through the cross-hairs and notes that the travelling microscope has moved a distance of 9.8×10^{-4} m.

The student uses this data to determine the thickness of the thin metal wire between the glass plates.

11. (continued)

(a) State whether the interference pattern is produced by division of amplitude or by division of wavefront. **1**

(b) Determine the diameter of the thin metal wire. **4**

Space for working and answer

(c) By measuring multiple fringe separations rather than just one, the student states that they have more confidence in the value of diameter of the wire which was obtained.

Suggest one reason why the student's statement is correct. **1**

(d) A current is now passed through the thin metal wire and its temperature increases.

The fringes are observed to get closer together.

Suggest a possible explanation for this observation. **2**

Page twenty-nine [Turn over

12. A student is observing the effect of passing light through polarising filters.

Two polarising filters, the polariser and the analyser, are placed between a lamp and the student as shown in Figure 12A.

The polariser is held in a fixed position, and the analyser can be rotated. Angle θ is the angle between the transmission axes of the two filters.

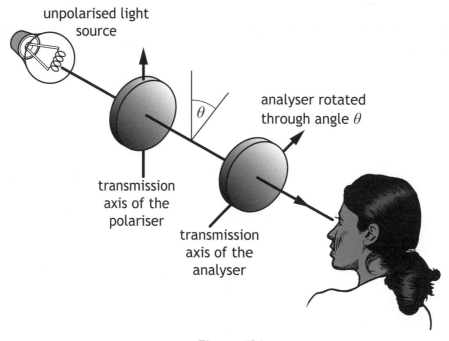

Figure 12A

When the transmission axes of the polariser and the analyser are parallel, θ is 0° and the student observes bright light from the lamp.

(a) (i) Describe, in terms of brightness, what the student observes as the analyser is slowly rotated from 0° to 180°. **2**

(ii) The polariser is now removed.

Describe, in terms of brightness, what the student observes as the analyser is again slowly rotated from 0° to 180° **1**

12. (continued)

(b) Sunlight reflected from a wet road can cause glare, which is hazardous for drivers. This is shown in Figure 12B

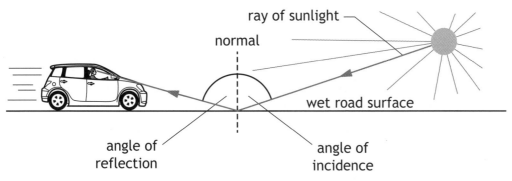

Figure 12B

Reflected sunlight is polarised when the light is incident on the wet road surface at the Brewster angle.

(i) Calculate the Brewster angle for light reflected from water. **3**

Space for working and answer

(ii) A driver is wearing polarising sunglasses.

Explain how wearing polarising sunglasses rather than non-polarising sunglasses will reduce the glare experienced by the driver. **1**

13. (a) State what is meant by *electric field strength*. **1**

(b) Two identical spheres, each with a charge of +22 nC, are suspended from point P by two equal lengths of light insulating thread.

The spheres repel and come to rest in the positions shown in Figure 13A.

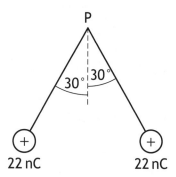

Figure 13A

(i) Each sphere has a weight of 9.80×10^{-4} N.

By considering the forces acting on one of the spheres, show that the electric force between the charges is 5.66×10^{-4} N. **2**

Space for working and answer

13. (b) (continued)

(ii) By considering the electric force between the charges, calculate the distance between the centres of the spheres. **3**

Space for working and answer

(iii) Calculate the electrical potential at point P due to both charged spheres. **5**

Space for working and answer

14. A student carries out an experiment to determine the charge to mass ratio of the electron.

    The apparatus is set up as shown in Figure 14A.

    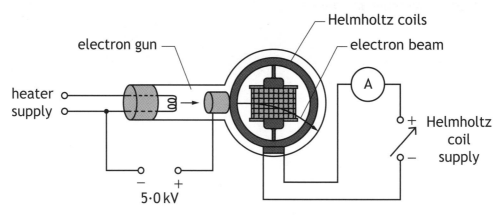

    Figure 14A

    An electron beam is produced using an electron gun connected to a 5·0 kV supply. A current I in the Helmholtz coils produces a uniform magnetic field.

    The electron beam enters the magnetic field.

    The path of the electron beam between points O and P can be considered to be an arc of a circle of constant radius r. This is shown in Figure 14B.

    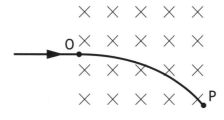

    Figure 14B

    The student records the following measurements:

Electron gun supply voltage, V	5·0 kV (±10%)
Current in the Helmholtz coils, I	0·22 A (±5%)
Radius of curvature of the path of the electron beam between O and P, r	0·28 m (±6%)

14. (continued)

    (a) The manufacturer's instruction sheet states that the magnetic field strength B at the centre of the apparatus is given by

    $$B = 4 \cdot 20 \times 10^{-3} \times I$$

    Calculate the magnitude of the magnetic field strength in the centre of the apparatus.

    Space for working and answer

    (b) The charge to mass ratio of the electron is calculated using the following relationship

    $$\frac{q}{m} = \frac{2V}{B^2 r^2}$$

    (i) Using the measurements recorded by the student, calculate the charge to mass ratio of the electron.

    Space for working and answer

    (ii) Determine the absolute uncertainty in the charge to mass ratio of the electron.

    Space for working and answer

14. (continued)

(c) A second student uses the same equipment to find the charge to mass ratio of the electron and analyses their measurements differently.

The current in the Helmholtz coils is varied to give a range of values for magnetic field strength. This produces a corresponding range of measurements of the radius of curvature.

The student then draws a graph and uses the gradient of the line of best fit to determine the charge to mass ratio of the electron.

Suggest which quantities the student chose for the axes of the graph. **1**

14. (continued)

(d) The graphical method of analysis used by the second student should give a more reliable value for the charge to mass ratio of the electron than the value obtained by the first student.

Use your knowledge of experimental physics to explain why this is the case.

3

15. A defibrillator is a device that gives an electric shock to a person whose heart has stopped beating normally.

This is shown in Figure 15A.

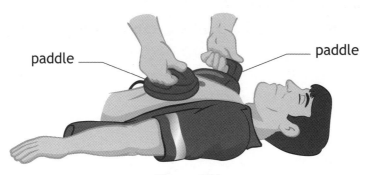

Figure 15A

Two paddles are initially placed in contact with the patient's chest.

A simplified defibrillator circuit is shown in Figure 15B.

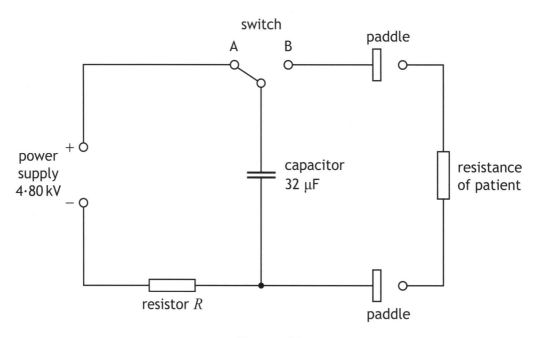

Figure 15B

When the switch is in position A, the capacitor is charged until there is a large potential difference across the capacitor.

15. (continued)

(a) The capacitor can be considered to be fully charged after 5 time constants.

The time taken for the capacitor to be considered to be fully charged is 10·0 s.

Determine the resistance of resistor R. **3**

Space for working and answer

(b) During a test, an 80·0 Ω resistor is used in place of the patient.

The switch is moved to position B, and the capacitor discharges through the 80·0 Ω resistor.

The initial discharge current is 60 A.

The current in the resistor will fall to half of its initial value after 0·7 time constants.

Show that the current falls to 30 A in 1·8 ms. **2**

Space for working and answer

15. (continued)

(c) In practice a current greater than 30 A is required for a minimum of 5·0 ms to force the heart of a patient to beat normally.

An inductor, of negligible resistance, is included in the circuit to increase the discharge time of the capacitor to a minimum of 5·0 ms.

This is shown in Figure 15C.

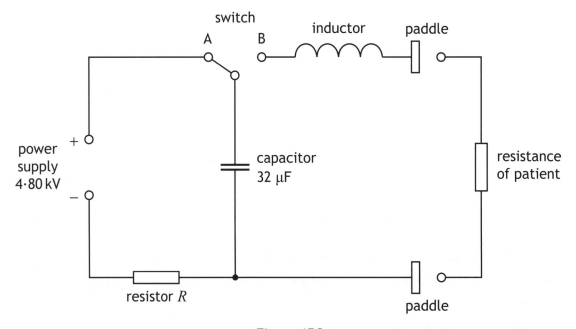

Figure 15C

(i) The inductor has an inductance of 50·3 mH.

The capacitor is again fully charged. The switch is then moved to position B.

Calculate the rate of change of current at the instant the switch is moved to position B.

Space for working and answer

15. (c) (continued)

(ii) It would be possible to increase the discharge time of the capacitor with an additional resistor connected in the circuit in place of the inductor. However, the use of an additional resistor would mean that maximum energy was not delivered to the patient.

Explain why it is more effective to use an inductor, rather than an additional resistor, to ensure that maximum energy is delivered to the patient. **2**

[END OF QUESTION PAPER]

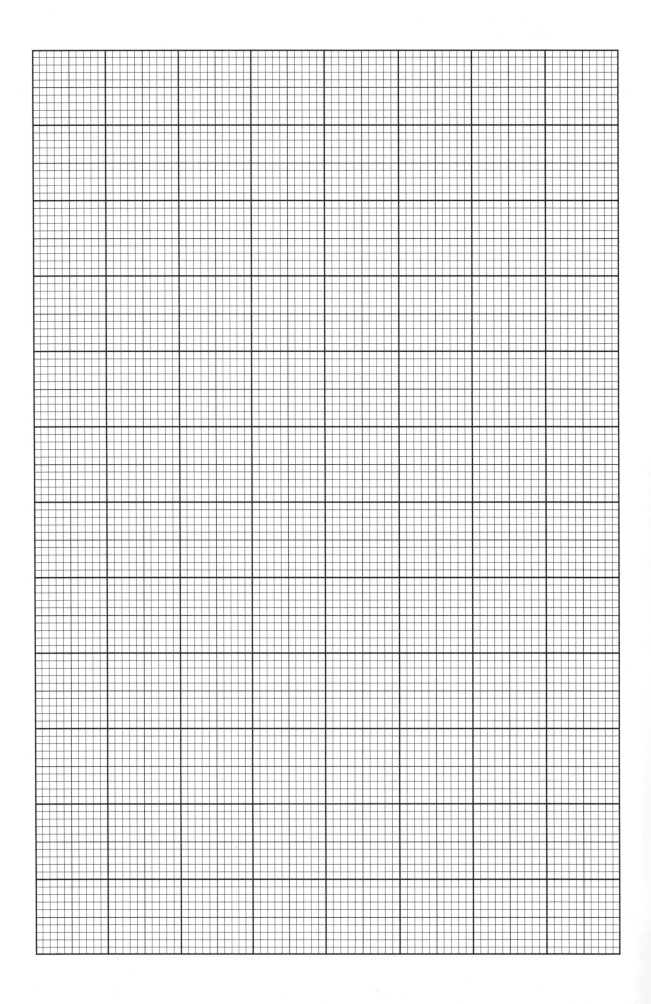

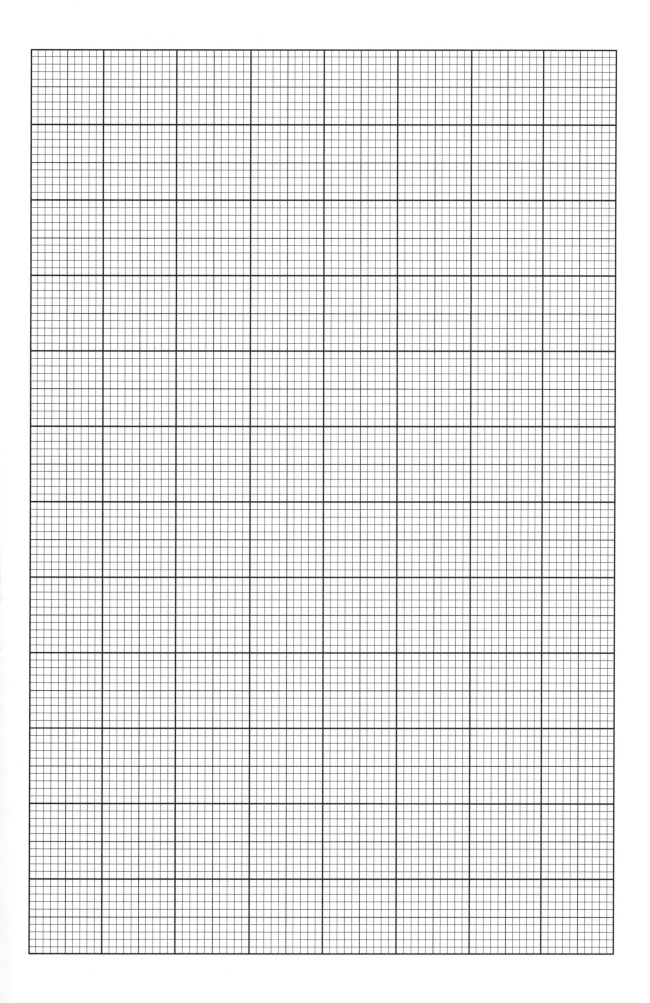

ADDITIONAL SPACE FOR ANSWERS AND ROUGH WORK

ADVANCED HIGHER

Answers

ANSWERS FOR SQA ADVANCED HIGHER PHYSICS 2018

ADVANCED HIGHER PHYSICS 2016

Question			Answer	Max mark
1.	(a)		$v = 0.135t^2 + 1.26t$ $a = \dfrac{dv}{dt} = 0.135 \times 2t + 1.26$ 1 $a = (0.135 \times 2 \times 15.0) + 1.26$ 1 $a = 5.31 \text{ ms}^{-2}$ 1	3
	(b)		$v = 0.135t^2 + 1.26t$ $s = \int_0^{150} v \cdot dt = [0.045t^3 + 0.63t^2]_0^{150}$ 1 $s = (0.045 \times 15.0^3) + (0.63 \times 15.0^2)$ 1 $s = 294$ m 1	3
2.	(a)	(i)	Velocity changing Or changing direction Or an unbalanced force is acting Or a centripetal/central/radial force is acting	1
		(ii)	Towards the centre	1
	(b)	(i) (A)	SHOW QUESTION $\omega = \dfrac{d\theta}{dt}$ OR $\omega = \dfrac{\theta}{t}$ 1 $\omega = \dfrac{1.5 \times 2\pi}{2.69}$ 1 $\omega = 3.5$ rad s^{-1}	2
		(B)	$F = mr\omega^2$ 1 $F = 0.059 \times 0.48 \times 3.5^2$ 1 $F = 0.35$ N 1	3
		(C)	$W = mg$ $W = 0.059 \times 9.8$ 1 $T^2 = 0.35^2 + (0.059 \times 9.8)^2$ 1 $T = 0.68$ N 1	3
		(ii)	In a straight line at a tangent to the circle	1

Question			Answer	Max mark
3.	(a)		$v = \sqrt{\dfrac{2GM}{r}}$ 1 $v = \sqrt{\dfrac{2 \times 6.67 \times 10^{-11} \times 9.5 \times 10^{12}}{2.1 \times 10^3}}$ 1 $v = 0.78$ (m s^{-1}) 1 (Lander returns to surface as) lander v less than escape velocity of comet 1	4
	(b)	(i)	SHOW QUESTION ($F_g = W$) $\dfrac{GMm}{r^2} = mg$ 1 for both eqns, 1 for equating $g = \dfrac{GM}{r^2}$ $g = \dfrac{6.67 \times 10^{-11} \times 9.5 \times 10^{12}}{(2.1 \times 10^3)^2}$ 1 $g = 1.4 \times 10^{-4}$ N kg^{-1}	3
		(ii)	Height will be greater 1 Because 'a' reduces 1 with height 1	3
4.	(a)		$b = \dfrac{L}{4\pi r^2}$ 1 $1.05 \times 10^{-9} = \dfrac{L}{4\pi (9.94 \times 10^{16})^2}$ 1 $L = 1.30 \times 10^{26}$ W 1	3
	(b)		$L = 4\pi r^2 \sigma T^4$ 1 $1.30 \times 10^{26} = 4\pi (5.10 \times 10^8)^2 \times 5.67 \times 10^{-8} \times T^4$ 1 $T = 5150$ K 1	3
	(c)		That the star is a black body (emitter/radiator) OR The star is spherical/constant radius OR The surface temperature of the star is constant/uniform OR No energy absorbed between star and Earth	1

Question			Answer	Max mark
5.	(a)	(i)	Frames of reference that are accelerating (with respect to an inertial frame)	1
		(ii)	It is impossible to tell the difference between the effects of gravity and acceleration.	1
	(b)	(i)	Any convex upward parabola	1
		(ii)	Any straight line	1
	(c)		The clock on the surface of the Earth would run more slowly. 1 The (effective) gravitational field for the spacecraft is smaller. 1 OR vice versa	2
6.			This is an open-ended question. Demonstrates no understanding 0 marks Demonstrates limited understanding 1 mark Demonstrates reasonable understanding 2 marks Demonstrates good understanding 3 marks **1 mark:** The student has demonstrated a limited understanding of the physics involved. The student has made some statement(s) which is/are relevant to the situation, showing that at least a little of the physics within the problem is understood.	3

Question			Answer	Max mark
			2 marks: The student has demonstrated a reasonable understanding of the physics involved. The student makes some statement(s) which is/are relevant to the situation, showing that the problem is understood. **3 marks:** The maximum available mark would be awarded to a student who has demonstrated a good understanding of the physics involved. The student shows a good comprehension of the physics of the situation and has provided a logically correct answer to the question posed. This type of response might include a statement of the principles involved, a relationship or an equation, and the application of these to respond to the problem. This does not mean the answer has to be what might be termed an "excellent" answer or a "complete" one.	
7.	(a)	(i)	$T_K = 15 + 273$ 1 $T_{kelvin} = \dfrac{b}{\lambda_{peak}}$ $288 = \dfrac{2 \cdot 89 \times 10^{-3}}{\lambda_{peak}}$ 1 $\lambda_{peak} = 1 \cdot 0 \times 10^{-5}$ m 1	3
		(ii)	Infrared	1
	(b)	(i)	(Curve) A 1 Peak at shorter wavelength/higher frequency (as temperature is higher) 1 **OR** Higher/greater (peak) intensity (as greater energy) 1	2

Question			Answer	Max mark
		(ii)	Curve asymptotic to y-axis and decreasing with increased wavelength	1
8.	(a)	(i)	The uncertainty in the momentum (in the x-direction)	1
		(ii)	The precise position of a particle/system and its momentum cannot both be known at the same instant. **1** **OR** If the uncertainty in the location of the particle is reduced, the minimum uncertainty in the momentum of the particle will increase (or vice-versa). **1** **OR** The precise energy and lifetime of a particle cannot both be known at the same instant. **1** **OR** If the uncertainty in the energy of the particle is reduced, the minimum uncertainty in the lifetime of the particle will increase (or vice-versa). **1**	1
	(b)	(i)	$\lambda = \dfrac{h}{p}$ **1** $\lambda = \dfrac{6\cdot 63 \times 10^{-34}}{6\cdot 5 \times 10^{-24}}$ **1** $\lambda = 1\cdot 0 \times 10^{-10}$ (m) **1** slit width $0\cdot 1$ nm used **1**	4
		(ii)	$\Delta x \, \Delta p_x \geq \dfrac{h}{4\pi}$ **1** $\Delta x \times 6\cdot 5 \times 10^{-26} \geq \dfrac{6\cdot 63 \times 10^{-34}}{4\pi}$ **1** $\Delta x \geq 8\cdot 1 \times 10^{-10}$ min uncertainty $= 8\cdot 1 \times 10^{-10}$ m **1**	3

Question			Answer	Max mark
	(b)	(iii)	Electron behaves like a wave "Electron shows interference" Uncertainty in position is greater than slit separation Electron passes through both slits Any three of the statements can be awarded 1 mark each	3
9.	(a)		SHOW QUESTION $m\dfrac{v^2}{r} = Bqv$ 1 for both relationships 1 for equating $r = \dfrac{mv}{Bq}$	2
	(b)	(i)	$1\cdot 50$ (MeV) $= 1\cdot 50 \times 10^6 \times 1\cdot 60 \times 10^{-19}$ $= 2\cdot 40 \times 10^{-13}$ (J) **1** $E_k = \dfrac{1}{2}mv^2$ **1** $2\cdot 40 \times 10^{-13} = 0\cdot 5 \times 3\cdot 34 \times 10^{-27} \times v^2$ **1** $v = 1\cdot 20 \times 10^7$ ms^{-1} **1**	4
		(ii)	$r = \dfrac{mv}{Bq}$ $2\cdot 50 = \dfrac{3\cdot 34 \times 10^{-27} \times 1\cdot 20 \times 10^7}{B \times 1\cdot 60 \times 10^{-19}}$ **1** $B = 0\cdot 100$ T **1**	2
		(iii)	r will be less **1** $r \propto \dfrac{m}{q}$ and q increases more than m does or q doubles but m increases by a factor of $1\cdot 5$ **1**	2
10.	(a)	(i)	Displacement is proportional to and in the opposite direction to the acceleration.	1

Question			Answer	Max mark
		(ii)	SHOW QUESTION $y = A\cos \omega t$ $\frac{dy}{dt} = -\omega A \sin \omega t$ $\frac{d^2y}{dt^2} = -\omega^2 A \cos \omega t$ — 1 $\frac{d^2y}{dt^2} = -\omega^2 y$ — 1 $\frac{d^2y}{dt^2} + \omega^2 y = 0$	2
	(b)	(i)	SHOW QUESTION $T = \frac{12 \cdot 0}{10}$ — 1 $\omega = \frac{2\pi}{T}$ — 1 $\omega = \frac{2\pi \times 10}{12}$ — 1 $\omega = 5 \cdot 2$ rad s^{-1}	3
		(ii)	$v = (\pm)\omega\sqrt{A^2 - y^2}$ — 1 $v = 5 \cdot 2 \times 0 \cdot 04$ — 1 $v = 0 \cdot 21$ m s^{-1} — 1	3
		(iii)	$E_P = \frac{1}{2}m\omega^2 y^2$ — 1 $E_P = \frac{1}{2} \times 1 \cdot 5 \times 5 \cdot 2^2 \times 0 \cdot 04^2$ — 1 $E_P = 0 \cdot 032$ J — 1	3
	(c)	(i)	Any valid method of damping	1
		(ii)	Amplitude of harmonic wave reducing	1
11.	(a)		$\frac{1}{\lambda} = 0 \cdot 357$ — 1 $\lambda = \frac{1}{0 \cdot 357}$ $v = f\lambda$ — 1 $v = 118 \times \frac{1}{0 \cdot 357}$ — 1 $v = 331$ m s^{-1} — 1	4
	(b)		$E = kA^2$ — 1 $\frac{E_1}{A_1^2} = \frac{E_2}{A_2^2}$ $\frac{1}{0 \cdot 250^2} = \frac{0 \cdot 5}{A_2^2}$ — 1 $A_2 = 0 \cdot 177$ (m) — 1 $\Delta y = 0 \cdot 177 \sin 2\pi (118t + 0 \cdot 357x)$ — 1	4
12.	(a)		(The axes should be arranged) at 90° to each other (e.g. horizontal and vertical).	1
	(b)		The filter for each eye will allow light from one projected image to pass through — 1 while blocking the light from the other projector. — 1	2
	(c)		There will be no change to the brightness. — 1 Light from the lamp is unpolarised. — 1	2
	(d)		(As the student rotates the filter,) the image from one projector will decrease in brightness, while the image from the other projector will increase in brightness. (The two images are almost identical.) — 1	1
13.	(a)		SHOW QUESTION $V = \frac{1}{4\pi\varepsilon_o}\frac{Q_1}{r}$ — 1 $V = \frac{1}{4\pi \times 8 \cdot 85 \times 10^{-12}} \frac{12 \times 10^{-9}}{0 \cdot 30}$ — 1 $V = (+)360$ V	2

ANSWERS FOR ADVANCED HIGHER PHYSICS

Question			Answer	Max mark
	(b)	(i)	$V = -360$ (V) 1 $\quad V = \dfrac{1}{4\pi\varepsilon_o} \dfrac{Q_2}{r}$ $\quad -360 = \dfrac{Q_2}{4\pi \times 8\cdot 85 \times 10^{-12} \times 0\cdot 40}$ 1 $\quad Q_2 = -1\cdot 6 \times 10^{-8}$ C 1	3
		(ii)	$E_1 = \dfrac{1}{4\pi\varepsilon_o} \dfrac{Q_1}{r^2}$ 1 $\quad E_1 = \dfrac{1}{4\pi \times 8\cdot 85 \times 10^{-12}} \dfrac{12 \times 10^{-9}}{0\cdot 30^2}$ 1 $\quad E_1 = 1200$ (N C^{-1} to right) $\quad E_2 = \dfrac{1}{4\pi \times 8\cdot 85 \times 10^{-12}} \dfrac{1\cdot 6 \times 10^{-8}}{0\cdot 40^2}$ 1 $\quad E_2 = 900$ (N C^{-1} to right) $\quad$ Total $= 2100$ N C^{-1} (to right) 1	4
		(iii)	Shape of attractive field, including correct direction 1 Skew in correct direction 1	2
14.	(a)		$B = \dfrac{\mu_o I}{2\pi r}$ 1 $\quad B = 5 \times 10^{-6} = \dfrac{4\pi \times 10^{-7} \times I}{2\pi \times 0\cdot 1}$ 1 $\quad I = 2\cdot 5$ A 1	3
	(b)	(i)	Ignore calibration (less than 1/3) $\quad \%u/c = \dfrac{0\cdot 002}{0\cdot 1} \times 100 = 2\%$ 1	1
		(ii)	Reading $5 = \dfrac{0\cdot 1}{5} \times 100 = 2\%$ 1 Total% $= \sqrt{(\text{reading}\%^2 + \text{calibration}\%^2)}$ 1 Total % $= \sqrt{(1\cdot 5^2 + 2^2)}$ $\qquad = 2\cdot 5\%$ 1	3

Question			Answer	Max mark
		(iii)	Total % $= \sqrt{(2^2 + 2\cdot 5^2)}$ $\qquad = \sqrt{10\cdot 25}\%$ 1 abs u/c $= \dfrac{\sqrt{10\cdot 25}}{100} \times 2\cdot 5$ 1 $\qquad = 0\cdot 08$ A	2
	(c)		Uncertainty in measuring exact distance from wire to position of sensor	1
15.	(a)	(i)	gradient $= \dfrac{8\cdot 3 \times 10^{-10}}{10^3}$ $\qquad\quad = 8\cdot 3 \times 10^{-13}$ 1 ----- gradient $= \varepsilon_0 A$ $8\cdot 3 \times 10^{-13} = \varepsilon_0 \times 9\cdot 0 \times 10^{-2}$ 1 $\varepsilon_0 = 9\cdot 2 \times 10^{-12}$ F m^{-1} 1	3
		(ii)	$c = \dfrac{1}{\sqrt{\varepsilon_0 \mu_0}}$ 1 $c = \dfrac{1}{\sqrt{9\cdot 2 \times 10^{-12} \times 4\pi \times 10^{-7}}}$ 1 $c = 2\cdot 9 \times 10^8$ m s^{-1} 1	3
	(b)		Systematic uncertainty specific to capacitance or spacing measurement	1
16.	(a)		$I = \dfrac{2}{5} mr^2$ 1 $I = \dfrac{2}{5} \times 3\cdot 8 \times 0\cdot 053^2$ 1 $I = 4\cdot 3 \times 10^{-3}$ kg m^2 1	3
	(b)	(i)	Labelling & scales 1 Plotting 1 Best fit line 1	3
		(ii)	gradient $= 1\cdot 73$ or consistent with candidate's best fit line 1 ----- $2gh = \left(\dfrac{I}{mr^2} + 1\right) v^2$ $\dfrac{2gh}{v^2} = \left(\dfrac{I}{mr^2} + 1\right)$ $1\cdot 73 = \left(\dfrac{I}{3\cdot 8 \times 0\cdot 053^2} + 1\right)$ 1 $I = 7\cdot 8 \times 10^{-3}$ kgm^2 1	3

Question		Answer	Max mark
	(c)	This is an open-ended question. Demonstrates no understanding 0 marks Demonstrates limited understanding 1 mark Demonstrates reasonable understanding 2 marks Demonstrates good understanding 3 marks **1 mark:** The student has demonstrated a limited understanding of the physics involved. The student has made some statement(s) which is/are relevant to the situation, showing that at least a little of the physics within the problem is understood. **2 marks:** The student has demonstrated a reasonable understanding of the physics involved. The student makes some statement(s) which is/are relevant to the situation, showing that the problem is understood. **3 marks:** The maximum available mark would be awarded to a student who has demonstrated a good understanding of the physics involved. The student shows a good comprehension of the physics of the situation and has provided a logically correct answer to the question posed.	3

Question	Answer	Max mark
	This type of response might include a statement of the principles involved, a relationship or an equation, and the application of these to respond to the problem. This does not mean the answer has to be what might be termed an "excellent" answer or a "complete" one.	

ADVANCED HIGHER PHYSICS 2017

Question			Answer	Max mark
1.	(a)		$v = 0\cdot4t^2 + 2t$ $v = (0\cdot4 \times 3\cdot10^2)$ $\quad + (2 \times 3\cdot10)$ 1 $v = 10\cdot0\text{ ms}^{-1}$ 1 Accept: 10, 10·04, 10·044	2
	(b)		$s = \int (0\cdot4t^2 + 2t)\,dt$ $s = \dfrac{0\cdot4}{3}t^3 + t^2(+c)$ 1 $s = 0$ when $t = 0$, $c = 0$ $s = \dfrac{0\cdot4}{3} \times (3\cdot10)^3 + 3\cdot10^2$ 1 $s = 13\cdot6\text{ m}$ 1 Accept: 14, 13·58, 13·582	3
2.	(a)	(i)	weight tension	1

ANSWERS FOR ADVANCED HIGHER PHYSICS

Question			Answer	Max mark
		(ii)	*Free body diagram showing a ball on a circular path with "tension" arrow pointing up and "weight" arrow pointing down*	1
	(iii) (A)		$T + (0.35 \times 9.8) = 4.0$ 1 $T = 0.57$ N 1 Accept: 0.6, 0.570, 0.5700	2
	(B)		$T = 7.4$ N 1 Accept: 7, 7.43, 7.430	1
(b)			the tension reduces (to zero) 1 weight is greater than the central force that would be required for circular motion. 1	2
(c)			Shape 1 0.48 and −0.48 for amplitude 1 0.7(0) time for half cycle 1 *Graph of s(m) vs t(s): sinusoidal curve starting at 0.48, crossing zero, reaching −0.48 at 0.70 s*	3
3.	(a)		$I = \frac{1}{2}mr^2$ 1 $I = \frac{1}{2} \times 0.40 \times (290 \times 10^{-3})^2$ 1 $I = 0.017$ kg m² 1 Accept: 0.02, 0.0168, 0.01682	3
	(b)	(i)	$T = Fr$ 1 $T = 8.0 \times 7.5 \times 10^{-3}$ 1 $T = 0.060$ Nm 1 Accept: 0.06, 0.0600, 0.06000	3

Question			Answer	Max mark
		(ii)	$T = I\alpha$ 1 $0.060 = 0.017 \times \alpha$ 1 $\alpha = 3.5$ rad s⁻² 1 Accept: 4, 3.53, 3.529	3
		(iii)	$\theta = \frac{s}{r}$ 1 $\theta = \frac{0.25}{7.5 \times 10^{-3}}$ 1 $\omega^2 = \omega_0^2 + 2\alpha\theta$ 1 $\omega^2 = 0^2 + 2 \times 3.5 \times \frac{0.25}{7.5 \times 10^{-3}}$ 1 $\omega = 15$ rad s⁻¹ 1 Accept: 20, 15.3, 15.28	5
	(c)		$I_{cube} = mr^2$ 1 $I_{cube} = 25 \times 10^{-3} \times (220 \times 10^{-3})^2$ 1 $I_1\omega_1 = (I_1 + I_{cube})\omega_2$ 1 $0.017 \times 12 = (0.017 + (25 \times 10^{-3} \times (220 \times 10^{-3})^2))\omega_2$ 1 $\omega_2 = 11$ rad s⁻¹ 1 Accept: 10, 11.2, 11.20	5
4.	(a)	(i)	$F = \frac{mv^2}{r}$ 1 $24.1 = \frac{1240 \times v^2}{(263 \times 10^3 + 680 \times 10^3)}$ 1 $v = 135$ m s⁻¹	2
		(ii)	$v_c = \frac{2\pi r}{T}$ 1 $135 = \frac{2\pi(263 \times 10^3 + 680 \times 10^3)}{T}$ 1 $T = 4.39 \times 10^4$ s 1 Accept: 4.4, 4.389, 4.3889	3
	(b)	(i)	The work done in moving unit mass from infinity (to that point). 1	1

ANSWERS FOR ADVANCED HIGHER PHYSICS

Question		Answer	Max mark
	(ii)	$V_{low} - V_{high} = -3 \cdot 22 \times 10^4$ $-(-1 \cdot 29 \times 10^4)$ **1** $V_{low} - V_{high} = -1 \cdot 93 \times 10^4$ $(\Delta)E = (\Delta)Vm$ **1** $(\Delta)E = -1 \cdot 93 \times 10^4 \times 1240$ **1** $(\Delta)E = -2 \cdot 39 \times 10^7$ J **1** Accept: $2 \cdot 4$, $2 \cdot 393$, $2 \cdot 3932$	4
5.		Demonstrates no understanding 0 marks Demonstrates limited understanding 1 mark Demonstrates reasonable understanding 2 marks Demonstrates good understanding 3 marks This is an open-ended question. **1 mark:** The student has demonstrated a limited understanding of the physics involved. The student has made some statement(s) which is/are relevant to the situation, showing that at least a little of the physics within the problem is understood.	3

Question		Answer	Max mark
		2 marks: The student has demonstrated a reasonable understanding of the physics involved. The student makes some statement(s) which is/are relevant to the situation, showing that the problem is understood. **3 marks:** The maximum available mark would be awarded to a student who has demonstrated a good understanding of the physics involved. The student shows a good comprehension of the physics of the situation and has provided a logically correct answer to the question posed. This type of response might include a statement of the principles involved, a relationship or an equation, and the application of these to respond to the problem. This does not mean the answer has to be what might be termed an "excellent" answer or a "complete" one.	
6.	(a)	(electron) neutrino (1) and positron (1)	2
	(b) (i)	Correctly marked 1	1
	(ii)	(White) Dwarf 1	1

ANSWERS FOR ADVANCED HIGHER PHYSICS

Question			Answer	Max mark
		(iii)	$L = 4.9 \times 10^{-4} \times 3.9 \times 10^{26}$ 1 $b = \dfrac{L}{4\pi r^2}$ 1 $1.3 \times 10^{-12} = \dfrac{4.9 \times 10^{-4} \times 3.9 \times 10^{26}}{4\pi r^2}$ 1 $r = 1.1 \times 10^{17}$ m 1 Accept: 1, 1.08, 1.082	4
	(c)		$\dfrac{L}{L_0} = 1.5\left(\dfrac{M}{M_0}\right)^{3.5}$ $\dfrac{L}{L_0} = 1.5\left(\dfrac{10.3}{1}\right)^{3.5}$ 1 $L = 5260(L_0)$ 1 Accept: 5300, 5260.4	2
7.	(a)		Atoms in the Nd:YAG have a shorter lifetime (in the excited state) OR Atoms in the Ar have a longer lifetime (in the excited state) 1 $\Delta f \propto \Delta E$ and $\Delta t \propto \dfrac{1}{\Delta E}$ or $\Delta t \propto \dfrac{1}{\Delta f}$ 1	2
	(b)	(i)	$\Delta E \Delta t \geq \dfrac{h}{4\pi}$ 1 $\Delta E_{(min)} \times 5.0 \times 10^{-6} = \dfrac{6.63 \times 10^{-34}}{4\pi}$ 1 $\Delta E_{(min)} = 1.1 \times 10^{-29}$ J 1 Accept: 1, 1.06, 1.055	3
		(ii)	$(\Delta)E = h(\Delta)f$ 1 $1.1 \times 10^{-29} = 6.63 \times 10^{-34} \times (\Delta)f$ 1 $(\Delta)f = 1.7 \times 10^4$ Hz 1 Accept: 2, 1.66, 1.659	3

Question			Answer	Max mark
8.	(a)		At $t = 0$ $\sin \omega t = 0$, which would mean that $y = 0$. This is not the case in the example here, where $y = A$ at $t = 0$ 1	1
	(b)	(i)	$(F =)(-)m\omega^2 y = (-)ky$ 1 $\omega^2 = \dfrac{ky}{my}$ $\omega = \sqrt{\dfrac{k}{m}}$ $\omega = 2\pi f$ 1 $2\pi f = \sqrt{\dfrac{k}{m}}$ $f = \dfrac{1}{2\pi}\sqrt{\dfrac{k}{m}}$	2
		(ii)	$f = \dfrac{1}{T} = \left(\dfrac{1}{0.80}\right)$ 1 $f = \dfrac{1}{2\pi}\sqrt{\dfrac{k}{m}}$ $\dfrac{1}{0.80} = \dfrac{1}{2\pi}\sqrt{\dfrac{k}{0.50}}$ 1 $k = 31$ N m^{-1} 1 Accept: 30, 30.8, 30.84	3
		(iii)	$T = \dfrac{0.80}{\sqrt{2}}$ 1 $T = 0.57$ s 1 Accept: 0.6, 0.566, 0.5657	2
	(c)		a, y, v vs time graphs (sinusoidal)	2
9.	(a)	(i)	$(\omega = 2\pi f)$ $922 = 2\pi f$ 1 $f = 147$ Hz	1

Question			Answer	Max mark
		(ii)	$4 \cdot 50 = \left(\dfrac{2\pi}{\lambda}\right)$ 1 $v = f\lambda$ 1 $v = 147 \times \left(\dfrac{2\pi}{4 \cdot 50}\right)$ 1 $v = 205 \text{ m s}^{-1}$ 1 Accept: 210, 205·3, 205·25	4
		(iii)	$E = kA^2$ 1 $\dfrac{E}{(9 \cdot 50 \times 10^{-4})^2} = \dfrac{E}{8 \times A^2}$ 1 $A = 3 \cdot 36 \times 10^{-4} \text{ m}$ 1 Accept: 3·4, 3·359, 3·3588	3
	(b)		$\mu = \dfrac{9 \cdot 0 \times 10^{-3}}{0 \cdot 69}$ 1 $v = \sqrt{\dfrac{T}{\mu}}$ $203 = \sqrt{\dfrac{T}{\dfrac{9 \cdot 0 \times 10^{-3}}{0 \cdot 69}}}$ 1 $T = 540 \text{ N}$ 1 Accept: 500, 538, 537·5	3
	(c)	(i)	Waves <u>reflected</u> from each end <u>interfere</u> (to create maxima and minima). 1	1
		(ii)	$f_3 = (3 \times 270 =) 810 \text{ Hz}$ 1	1
10.	(a)		Pattern produced by <u>interference</u>. 1 Slits <u>horizontal and vertical</u> or at <u>right angles</u> 1	2
	(b)		The spots are closer together. 1 The green light has a shorter wavelength and since $d\sin\theta = m\lambda$, d is fixed, $(\sin)\theta$ is smaller. 1	2

Question			Answer	Max mark
	(c)	(i)	$\Delta x = \dfrac{\lambda D}{d}$ 1 $13 \cdot 0 \times 10^{-3} = \dfrac{\lambda \times 8 \cdot 11}{0 \cdot 41 \times 10^{-3}}$ 1 $\lambda = 6 \cdot 6 \times 10^{-7} \text{ m}$ 1 Accept: 7, 6·57, 6·572	3
		(ii)	% Uncertainty in fringe separation $= \left(\dfrac{0 \cdot 5}{13 \cdot 0}\right) \times 100$ 1 $= 3 \cdot 85\%$ % Uncertainty in slit separation $= \left(\dfrac{0 \cdot 01}{0 \cdot 41}\right) \times 100$ 1 $= 2 \cdot 44\%$ % Uncertainty in slit-screen separation $= \left(\dfrac{0 \cdot 01}{8 \cdot 11}\right) \times 100$ 1 $= 0 \cdot 123\%$ (can be ignored) % uncertainty in wavelength $= \sqrt{\left(\dfrac{0 \cdot 5}{13 \cdot 0}\right)^2 + \left(\dfrac{0 \cdot 01}{0 \cdot 41}\right)^2} \times 100\%$ $= 4 \cdot 56\%$ 1 $\Delta\lambda = \dfrac{4 \cdot 56}{100} \times 6 \cdot 6 \times 10^{-7}$ $\Delta\lambda = 0 \cdot 3 \times 10^{-7} \text{ m}$ 1	5
		(iii)	Increasing the slit-screen distance spreads out the fringes, <u>reducing the (percentage) uncertainty in the fringe separation</u> (which is the dominant uncertainty). 1	1
11.	(a)		Force acting per unit positive charge (in an electric field) 1	1

ANSWERS FOR ADVANCED HIGHER PHYSICS

Question			Answer		Max mark
	(b)	(i)	$E = \dfrac{Q}{4\pi\varepsilon_0 r^2}$	1	3
			$E = \dfrac{4\cdot 0\times 10^{-9}}{4\pi\times 8\cdot 85\times 10^{-12}\times 0\cdot 12^2}$	1	
			$E_{total} = \sqrt{2\times\left[\dfrac{4\cdot 0\times 10^{-9}}{4\pi\times 8\cdot 85\times 10^{-12}\times 0\cdot 12^2}\right]^2}$	1	
			$E_{total} = 3\cdot 5\times 10^3 \text{ N C}^{-1}$		
		(ii)	$F = QE$	1	3
			$F = 1\cdot 9\times 10^{-9}\times 3\cdot 5\times 10^3$	1	
			$F = 6\cdot 7\times 10^{-6}$ N	1	
			Accept: 7, 6·65, 6·650		
		(iii)	Towards the top of the page.	1	1
		(iv)	$r = \sqrt{(0\cdot 12^2 + 0\cdot 12^2)}$	1	4
			$F = \dfrac{Q_1 Q_2}{4\pi\varepsilon_0 r^2}$	1	
			$6\cdot 7\times 10^{-6}$		
			$= \dfrac{1\cdot 9\times 10^{-9}\times Q_2}{4\pi\times 8\cdot 85\times 10^{-12}\times\sqrt{(0\cdot 12^2+0\cdot 12^2)}^2}$	1	
			$Q_2 = 1\cdot 1\times 10^{-8}$ C	1	
			Accept: 1, 1·13, 1·129		
12.	(a)		Towards the top of the page.	1	1
	(b)		$F = qvB$	1	3
			$4\cdot 00\times 10^{-15} = 1\cdot 60\times 10^{-19}$		
			$\times v\times 115\times 10^{-3}$	1	
			$v = 2\cdot 17\times 10^5$ m s^{-1}	1	
			Accept: 2·2, 2·174, 2·1739		
	(c)		(Since $F = Bqv$) At lower speeds the magnetic force is reduced.	1	2
			Therefore unbalanced force (or acceleration) is downwards Or The magnetic force is less than the electric force	1	
	(d)		(All undeflected ions travel at) $2\cdot 17\times 10^5$ m s^{-1}	1	2
			relative size of forces is independent of mass and of charge.	1	

Question			Answer		Max mark
13.	(a)		$t = RC$	1	2
			$t = 1\cdot 0\times 10^3\times 1\cdot 0$	1	
			$t = 1\cdot 0\times 10^3$ s		
	(b)	(i)	circuit diagram showing (12V) d.c. supply, resistor and capacitor all in series. Values not required. Voltmeter or CRO connected across the capacitor.	1	1
		(ii) (A)	(After 1 time constant or 1000 s) $V = 7\cdot 6\ (V)$	1	2
			$\dfrac{V_c}{V_s} = \dfrac{7\cdot 6}{12}$	1	
			$\dfrac{V_c}{V_s} = 63\%$		
		(ii) (B)	4·5 – 5	1	1
	(c)		Demonstrates no understanding 0 marks		3
			Demonstrates limited understanding 1 mark		
			Demonstrates reasonable understanding 2 marks		
			Demonstrates good understanding 3 marks		
			This is an open-ended question.		
			1 mark: The student has demonstrated a limited understanding of the physics involved. The student has made some statement(s) which is/are relevant to the situation, showing that at least a little of the physics within the problem is understood.		
			2 marks: The student has demonstrated a reasonable understanding of the physics involved. The student makes some statement(s) which is/are relevant to the situation, showing that the problem is understood.		

Question			Answer	Max mark
			3 marks: The maximum available mark would be awarded to a student who has demonstrated a good understanding of the physics involved. The student shows a good comprehension of the physics of the situation and has provided a logically correct answer to the question posed. This type of response might include a statement of the principles involved, a relationship or an equation, and the application of these to respond to the problem. This does not mean the answer has to be what might be termed an "excellent" answer or a "complete" one.	
14.	(a)	(i)	Capacitor has low reactance/impedance for high frequencies (therefore more current (and power) will be delivered to the tweeter at high frequencies). **1**	1
		(ii)	Inductor has low reactance/impedance for low frequencies (therefore more current (and power) will be delivered to the woofer at low frequencies). **1**	1
	(b)		$X_L = 2\pi f L$ **1** $8 \cdot 0 = 2 \times \pi \times 3 \cdot 0 \times 10^3 \times L$ **1** $L = 4 \cdot 2 \times 10^{-4}$ H **1** Accept: 4, 4·24, 4·244	3

Question			Answer	Max mark
	(c)	(i)	$\dfrac{dI}{dt} = 20 \cdot 0$ **1** $E = -L\dfrac{dI}{dt}$ **1** $-9 \cdot 0 = -L \times 20 \cdot 0$ **1** $L = 0 \cdot 45$ H **1** Accept: 0·5, 0·450, 0·4500	4
		(ii)	large (back) EMF. **1**	1

ADVANCED HIGHER PHYSICS 2018

Question			Answer	Max mark
1.	(a)		$v = 0 \cdot 0071t - 0 \cdot 00025t^2$ $a = \left(\dfrac{dv}{dt}\right) = 0 \cdot 0071 - 0 \cdot 0005t$ **1** $a = 0 \cdot 0071 - (0 \cdot 0005 \times 20 \cdot 0)$ **1** $a = -0 \cdot 0029$ m s^{-2} **1** Accept −0·003	3
	(b)		$v = 0 \cdot 0071t - 0 \cdot 00025t^2$ $s \left(= \displaystyle\int_0^{20 \cdot 0} v \cdot dt \right)$ $= \left[\dfrac{0 \cdot 0071}{2}t^2 - \dfrac{0 \cdot 00025}{3}t^3\right]_0^{20 \cdot 0}$ **1** $s = \left(\dfrac{0 \cdot 0071}{2} \times 20 \cdot 0^2\right)$ $\qquad - \left(\dfrac{0 \cdot 00025}{3} \times 20 \cdot 0^3\right) - 0$ **1** $s = 0 \cdot 75$ m **1** Accept 0·8, 0·753, 0·7533	3
2.	(a)	(i)	The car's direction/velocity is changing. **OR** Unbalanced/centripetal/central force acting on the car	1

ANSWERS FOR ADVANCED HIGHER PHYSICS

Question			Answer	Max mark
		(ii)	$a_{(r)} = \dfrac{v^2}{r}$ 1 $a_{(r)} = \dfrac{3 \cdot 5^2}{1 \cdot 8}$ 1 $a_{(r)} = 6 \cdot 8 \text{ ms}^{-2}$ 1 Accept: 7, 6·81, 6·806 Alternative acceptable response: $a_r = r\omega^2$ 1 $a_r = 1 \cdot 8 \times \left(\dfrac{3 \cdot 5}{1 \cdot 8}\right)^2$ 1 $a_r = 6 \cdot 8 \text{ ms}^{-2}$ 1	3
		(iii)	$F = \dfrac{mv^2}{r}$ $F = \dfrac{0 \cdot 431 \times 5 \cdot 5^2}{1 \cdot 8}$ 1 $F = 7 \cdot 2 \text{(N)}$ 1 Since 7·2(N) > 6·4(N) **OR** There is insufficient friction and the car does not stay on the track. 1	3
	(b)	(i)	$\left(F_{(\text{centripetal})} = \dfrac{mv^2_{(\text{max})}}{r}, W = mg\right)$ $\dfrac{mv^2_{(\text{max})}}{r} = mg$ 1, 1 $\dfrac{v^2_{(\text{max})}}{r} = g$ $v_{(\text{max})} = \sqrt{gr}$	2
		(ii)	$v_{(\text{max})} = \sqrt{gr}$ $v_{(\text{max})} = \sqrt{9 \cdot 8 \times 0 \cdot 65}$ 1 $v_{(\text{max})} = 2 \cdot 5 \text{ ms}^{-1}$ 1 Accept: 3, 2·52, 2·524	2
		(iii)	The second car will not lose contact with the track. 1 A smaller centripetal force is supplied by a smaller weight. 1	2

Question			Answer	Max mark
3.	(a)		$\omega = \omega_o + \alpha t$ 1 $\omega = 0 + (6 \cdot 7 \times 3 \cdot 9)$ 1 $\omega = 26 \text{ rads}^{-1}$ If final answer not shown 1 mark max	2
	(b)		$E_{(k)} = \dfrac{1}{2} I \omega^2$ 1 $430 = \dfrac{1}{2} \times I \times 26^2$ 1 $I = 1 \cdot 3 \text{ kg m}^2$ 1 Accept: 1, 1·27, 1·272	3
	(c)	(i)	$\theta = 14 \times 2\pi$ 1 $\omega^2 = \omega_o^2 + 2\alpha\theta$ 1 $0^2 = 26^2 + (2 \times \alpha \times 14 \times 2\pi)$ 1 $\alpha = -3 \cdot 8 \text{ rad s}^{-2}$ 1 Accept: −4, −3·84, −3·842	4
		(ii)	$T = I\alpha$ 1 $T = 1 \cdot 3 \times (-)3 \cdot 8$ 1 $T = (-)4 \cdot 9 \text{ Nm}$ 1 Accept: 5, 4·94, 4·940	3
4.	(a)	(i)	$(F_{\text{centripetal}} = F_{\text{gravitational}})$ 1 $mr\omega^2 = \dfrac{GMm}{r^2}$ 1 $\omega = \dfrac{2\pi}{T}$ or $\omega^2 = \left(\dfrac{2\pi}{T}\right)^2$ 1 $\dfrac{4\pi^2}{T^2} = \dfrac{GM}{r^3}$ $T^2 = \dfrac{4\pi^2}{GM} r^3$	3
		(ii)	$T^2 = \dfrac{4\pi^2}{GM} r^3$ $\left((197 \times 24 \times 60 \times 60)^2 = \dfrac{4\pi^2 \times (0 \cdot 63 \times 1 \cdot 5 \times 10^{11})^3}{6 \cdot 67 \times 10^{-11} \times M}\right)$ 1,1 $M = 1 \cdot 7 \times 10^{30} \text{(kg)}$ 1 Accept: 2, 1·72, 1·724	3

Question			Answer	Max mark
	(b)		$v = \sqrt{\dfrac{2GM}{r}}$ 1 $v = \sqrt{\dfrac{2 \times 6 \cdot 67 \times 10^{-11} \times 1 \cdot 7 \times 10^{30}}{1 \cdot 58 \times 10^{11}}}$ 1 $v = 3 \cdot 8 \times 10^4 \, \text{m s}^{-1}$ 1 Accept 4, 3·79, 3·789	3
5.	(a)		The Schwarzschild radius is the distance from the centre of a mass such that, the escape velocity at that distance would equal the speed of light. OR The Schwarzschild radius is the distance from the centre of a mass to the event horizon.	1
	(b)	(i)	$r_{(Schwarzschild)} = \dfrac{2GM}{c^2}$ 1 $r_{(Schwarzschild)} = \dfrac{2 \times 6 \cdot 67 \times 10^{-11} \times 2 \cdot 0 \times 10^{30}}{(3 \cdot 00 \times 10^8)^2}$ 1 $r_{(Schwarzschild)} = 3 \cdot 0 \times 10^3 \, \text{m}$ 1 Accept: 3×10^3, $2 \cdot 96 \times 10^3$, $2 \cdot 964 \times 10^3$	3
		(ii)	(Radius of Sun is $6 \cdot 955 \times 10^8$ m) This is greater than the Schwarzschild radius (the Sun is not a black hole.) 1	1
	(c)		$\phi = 3\pi \dfrac{r_s}{a(1-e^2)}$ $\phi = 3\pi \dfrac{3000}{5 \cdot 805 \times 10^{10} \times (1 - 0 \cdot 206^2)}$ 1 Angular change after one year = $4 \times \phi$ 1 Angular change = $2 \cdot 0 \times 10^{-6}$ rad 1 Accept 2, 2·03, 2·035 If 3·14 used, accept 2·034	3

Question		Answer	Max mark
6.		Demonstrates no understanding 0 marks Demonstrates limited understanding 1 mark Demonstrates reasonable understanding. 2 marks Demonstrates good understanding 3 marks This is an open-ended question. **1 mark:** The student has demonstrated a limited understanding of the physics involved. The student has made some statement(s) which is/are relevant to the situation, showing that at least a little of the physics within the problem is understood. **2 marks:** The student has demonstrated a reasonable understanding of the physics involved. The student makes some statement(s) which is/are relevant to the situation, showing that the problem is understood **3 marks:** The maximum available mark would be awarded to a student who has demonstrated a good understanding of the physics involved. The student shows a good comprehension of the physics of the situation and has provided a logically correct answer to the question posed. This type of response might include a statement of the principles involved, a relationship or an equation, and the application of these to respond to the problem. This does not mean the answer has to be what might be termed an "excellent" answer or a "complete" one.	3

Question			Answer	Max mark
7.	(a)	(i)	Neutrons are accelerated.	1
		(ii)	$\lambda = \dfrac{h}{p}$ 1 $\lambda = \dfrac{6 \cdot 63 \times 10^{-34}}{1 \cdot 29 \times 10^{-23}}$ 1 $\lambda = 5 \cdot 14 \times 10^{-11}$ m 1 Accept 5·1, 5·140, 5·1395	3
		(iii)	The <u>precise/exact</u> position of a particle and its momentum cannot both be known <u>at the same instant</u>. 1 OR If the (minimum) uncertainty in the position of a particle is reduced, the uncertainty in the momentum of the particle will increase (or vice-versa). 1	1
		(iv)	$\Delta p_x = p \times \dfrac{\%p}{100}$ $\Delta p_x = 1 \cdot 29 \times 10^{-23} \times \dfrac{3}{100}$ 1 $\Delta x_{min} \, \Delta p_x = \dfrac{h}{4\pi}$ or $\Delta x \, \Delta p_x \geq \dfrac{h}{4\pi}$ 1 $\Delta x_{(min)}$ $= \dfrac{6 \cdot 63 \times 10^{-34}}{4\pi \times 1 \cdot 29 \times 10^{-23} \times 0 \cdot 03}$ 1 $\Delta x_{(min)} = 1 \cdot 36 \times 10^{-10}$ m 1 Accept 1·4, 1·363, 1·3633	4
	(b)		The uncertainty in position will be (too) small. 1 Neutrons can be considered a particle/ cannot be considered a wave, even on the length scale of the lattice spacing. 1	2

Question			Answer	Max mark
8.	(a)	(i)	Main sequence	1
		(ii)	X: helium (nucleus) 1 Z: positron 1 Accept alpha particle (for particle X) Accept anti-electron (for particle Z) Accept He, e^+, β^+ (for both particles).	2
	(b)		Solar wind	1
	(c)	(i)	$p = 1 \cdot 6726 \times 10^{-6} \times n \times v^2$ $0 \cdot 956 = 1 \cdot 6726 \times 10^{-6} \times n \times 602^2$ 1 $n = 1 \cdot 58$ (particles per cm^3) 1 Accept 1·6, 1·577, 1·5771	2
		(ii) (A)	(As the particles are ejected in all directions they will) spread out (as they get further from the Sun).	1
		(ii) (B)	(The particles lose kinetic energy and) gain (gravitational) potential (energy) (as they move further from the Sun.) OR Work is done against the Sun's gravitational field (for the particles to move away). Accept reduction in velocity due to gravitational force and statement of $E_K = \dfrac{1}{2} mv^2$	1
	(d)		The charged particles have a component (of velocity) parallel to the (magnetic) field which moves them forwards in that direction. 1 The component (of velocity) perpendicular to the (magnetic) field causes a central force on the charged particle OR it moves in a circle. 1	2

ANSWERS FOR ADVANCED HIGHER PHYSICS

Question			Answer	Max mark
9.	(a)	(i)	$\omega = \dfrac{d\theta}{dt}$ 1 $\omega = \dfrac{2\pi \times 1 \cdot 5}{2 \cdot 5}$ 1 $\omega = 3 \cdot 8 \text{ rad s}^{-1}$ Accept $\omega = \dfrac{\theta}{t}$, $\omega = 2\pi f$ or $\omega = \dfrac{2\pi}{T}$ as a starting point.	2
		(ii)	$(x = -0 \cdot 2\cos(3 \cdot 8t))$ $\dfrac{dx}{dt} = -3 \cdot 8 \times (-0 \cdot 2\sin(3 \cdot 8t))$ $\dfrac{d^2x}{dt^2} = -3 \cdot 8^2 \times (-0 \cdot 2\cos(3 \cdot 8t))$ 1 $\dfrac{d^2x}{dt^2} = -3 \cdot 8^2 x$ 1 (Since the equation is in the form $a = -\omega^2 y$ or $a = -\omega^2 x$ the horizontal displacement is consistent with SHM). 1	3
		(iii)	$v = (\pm)\omega\sqrt{(A^2 - y^2)}$ 1 $v = (\pm)3 \cdot 8 \times \sqrt{(0 \cdot 2^2 - 0^2)}$ 1 $v = (\pm)0 \cdot 76 \text{ m s}^{-1}$ 1 Accept $v_{(max)} = (\pm)\omega A$ Accept $A = 0 \cdot 2$ m or $A = -0 \cdot 2$ m Accept $\dfrac{dx}{dt} = -3 \cdot 8 \times (-0 \cdot 2\sin(3 \cdot 8t))$ as a starting point. Accept 0·8, 0·760, 0·7600	3
		(iv)	$\dfrac{1}{2}(m)v^2 = (m)gh$ 1 $h = \dfrac{0 \cdot 5 \times 0 \cdot 76^2}{9 \cdot 8}$ 1 $h = 2 \cdot 9 \times 10^{-2}$ m 1 Accept 3, 2·95, 2·947	3

Question			Answer	Max mark
	(b)		The shape of the line should resemble a sinusoidal wave with values either all positive or all negative and the minimum vertical displacement consistent. 1 Peak height should show a steady decline with each oscillation/decreasing amplitude. 1	2
10.	(a)	(i)	$\phi = \dfrac{2\pi x}{\lambda}$ 1 $\phi = \dfrac{2\pi \times 4 \cdot 25 \times 10^{-7}}{1 \cdot 55 \times 10^{-6}}$ 1 $\phi = 1 \cdot 72$ rad 1 Accept 1·7, 1·723, 1·7228	3
		(ii)	(The electric field vectors will be in) <u>opposite</u> (directions at positions P and Q).	1
	(b)	(i)	$v = f\lambda$ 1 $v = 1 \cdot 31 \times 10^{14} \times 1 \cdot 55 \times 10^{-6}$ 1 $v = 2 \cdot 03 \times 10^8$ m s^{-1}	2
		(ii)	$v_m = \dfrac{1}{\sqrt{\varepsilon_m \mu_m}}$ $2 \cdot 03 \times 10^8 = \dfrac{1}{\sqrt{\varepsilon_m \times 4\pi \times 10^{-7}}}$ 1 $\varepsilon_m = 1 \cdot 93 \times 10^{-11}$ F m^{-1} 1 Accept 1·9, 1·931, 1·9311	2

ANSWERS FOR ADVANCED HIGHER PHYSICS

Question			Answer	Max mark
11.	(a)		(Division of) amplitude	1
	(b)		$\Delta x = \dfrac{9 \cdot 8 \times 10^{-4}}{20}$ 1 $\Delta x = \dfrac{\lambda l}{2d}$ 1 $d = \dfrac{589 \times 10^{-9} \times 75 \times 10^{-3} \times 20}{2 \times 9 \cdot 8 \times 10^{-4}}$ 1 $d = 4 \cdot 5 \times 10^{-4}$ m 1 Accept 5, 4·51, 4·508	4
	(c)		Reduces the uncertainty in the value of Δx or d obtained. **OR** Reduces the impact/significance of any uncertainty on the value obtained for Δx or d.	1
	(d)		The wire expands/d increases 1 $\Delta x = \dfrac{\lambda l}{2d}$, (and since d increases) while l and λ remain constant, (Δx decreases). **OR** (Since d increases and) $\Delta x \propto \dfrac{1}{d}$, ($\Delta x$ decreases). 1	2
12.	(a)	(i)	The brightness (starts at a maximum and) decreases to (a minimum at) 90°. 1 The brightness then increases (from the minimum back to the maximum at 180°). 1	2
		(ii)	The brightness remains constant (throughout).	1
	(b)	(i)	$n = \tan i_p$ 1 $i_p = \tan^{-1}(1 \cdot 33)$ 1 $i_p = 53 \cdot 1°$ 1 Accept 53, 53·06, 53·061	3
		(ii)	The polarising sunglasses will act as an analyser/absorb/block (some of) the glare.	1

Question			Answer	Max mark
13.	(a)		Force per unit positive charge (at a point in an electric field)	1
	(b)	(i)	$F_e = W \tan \theta$ 1 $F_e = 9 \cdot 80 \times 10^{-4} \times \tan 30$ 1 $F_e = 5 \cdot 66 \times 10^{-4}$ N	2
		(ii)	$F = \dfrac{Q_1 Q_2}{4 \pi \varepsilon_0 r^2}$ 1 $5 \cdot 66 \times 10^{-4} = \dfrac{(22 \times 10^{-9})^2}{4 \pi \times 8 \cdot 85 \times 10^{-12} r^2}$ 1 $r = 0 \cdot 088$ m 1 Accept 0·09, 0·0877, 0·08769 Accept 0·08773 if 9×10^9 used.	3
		(iii)	$V = \dfrac{Q}{4 \pi \varepsilon_0 r}$ 1 $r = 0 \cdot 088$ (m) 1 $V = \dfrac{22 \times 10^{-9}}{4 \times \pi \times 8 \cdot 85 \times 10^{-12} \times 0 \cdot 088}$ 1 $V_{total} = 2 \times \dfrac{22 \times 10^{-9}}{4 \times \pi \times 8 \cdot 85 \times 10^{-12} \times 0 \cdot 088}$ 1 $V_{total} = 4 \cdot 5 \times 10^3$ V 1 Accept: 4000, 4496	5
14.	(a)		$B = 4 \cdot 2 \times 10^{-3} \times 0 \cdot 22$ 1 $= 9 \cdot 2 \times 10^{-4}$ T 1 Accept 9, 9·24, 9·240	1
	(b)	(i)	$\dfrac{q}{m} = \dfrac{2V}{B^2 r^2}$ $\dfrac{q}{m} = \dfrac{2 \times 5 \cdot 0 \times 10^3}{(9 \cdot 2 \times 10^{-4})^2 \times 0 \cdot 28^2}$ 1 $\dfrac{q}{m} = 1 \cdot 5 \times 10^{11}$ C kg^{-1} 1 Accept 2, 1·51, 1·507	2
		(ii)	%Uncertainty in B & r is doubled 1 $\%\Delta(w) = \sqrt{(\%\Delta x^2 + \%\Delta y^2 + \%\Delta z^2)}$ 1 $\%\Delta\left(\dfrac{q}{m}\right) = \sqrt{(10^2 + 10^2 + 1 \cdot 2^2)}$ 1 $\Delta\left(\dfrac{q}{m}\right) = 0 \cdot 3 \times 10^{11}$ C kg^{-1} 1	4

Question		Answer	Max mark
	(c)	B^2 and $1/r^2$ (r^2 and $1/B^2$) OR B and $1/r$ (r and $1/B$) OR I and $1/r$ (r and $1/I$) OR I^2 and $1/r^2$ (r^2 and $1/I^2$)	1
	(d)	Demonstrates no understanding 0 marks Demonstrates limited understanding 1 mark Demonstrates reasonable understanding 2 marks Demonstrates good understanding. 3 marks This is an open-ended question. **1 mark:** The student has demonstrated a limited understanding of the physics involved. The student has made some statement(s) which is/are relevant to the situation, showing that at least a little of the physics within the problem is understood. **2 marks:** The student has demonstrated a reasonable understanding of the physics involved. The student makes some statement(s) which is/are relevant to the situation, showing that the problem is understood. **3 marks:** The maximum available mark would be awarded to a student who has demonstrated a good understanding of the physics involved. The student shows a good comprehension of the physics of the situation and has provided a logically correct answer to the question posed.	3

Question			Answer	Max mark
			This type of response might include a statement of the principles involved, a relationship or an equation, and the application of these to respond to the problem. This does not mean the answer has to be what might be termed an "excellent" answer or a "complete" one.	
15.	(a)		$t = RC$ 1 $\dfrac{10 \cdot 0}{5} = R \times 32 \times 10^{-6}$ 1 $R = 6 \cdot 3 \times 10^4 \, \Omega$ 1 Accept 6, 6·25, 6·250	3
	(b)		$t = RC$ 1 $t_{(\frac{1}{2})} = 0 \cdot 7 \times 80 \cdot 0 \times 32 \times 10^{-6}$ 1 $t_{(\frac{1}{2})} = 1 \cdot 8 \times 10^{-3} \, s$	2
	(c)	(i)	$\varepsilon = -L \dfrac{dI}{dt}$ 1 $-4 \cdot 80 \times 10^3 = -50 \cdot 3 \times 10^{-3} \times \dfrac{dI}{dt}$ 1 $\dfrac{dI}{dt} = 9 \cdot 54 \times 10^4 \, A\,s^{-1}$ 1 Accept 9·5, 9·543, 9·5427	3
		(ii)	(Additional) resistor will dissipate energy. 1 Inductor will store energy (and then deliver it to the patient). 1	2